STREPTOMYCES
in Nature and Medicine

STREPTOMYCES
in Nature and Medicine

The Antibiotic Makers

David A. Hopwood

John Innes Centre

OXFORD
UNIVERSITY PRESS

2007

OXFORD
UNIVERSITY PRESS

Oxford University Press, Inc., publishes works that further
Oxford University's objective of excellence
in research, scholarship, and education.

Oxford   New York
Auckland   Cape Town   Dar es Salaam   Hong Kong   Karachi
Kuala Lumpur   Madrid   Melbourne   Mexico City   Nairobi
New Delhi   Shanghai   Taipei   Toronto

With offices in
Argentina   Austria   Brazil   Chile   Czech Republic   France   Greece
Guatemala   Hungary   Italy   Japan   Poland   Portugal   Singapore
South Korea   Switzerland   Thailand   Turkey   Ukraine   Vietnam

Copyright © 2007 by Oxford University Press, Inc.

Published by Oxford University Press, Inc.
198 Madison Avenue, New York, New York 10016

www.oup.com

Oxford is a registered trademark of Oxford University Press.

All rights reserved. No part of this publication may be reproduced,
stored in a retrieval system, or transmitted, in any form or by any means,
electronic, mechanical, photocopying, recording, or otherwise,
without the prior permission of Oxford University Press.

Library of Congress Cataloging-in-Publication Data
Hopwood, D. A.
   Streptomyces in nature and medicine / David A. Hopwood.
      p. ; cm.
   Includes bibliographical references and index.
   ISBN-13 978–0–19–515066–7
   1. Streptomyces—Genetics.
   [DNLM: 1. Streptomyces—genetics.   2. Genetic Engineering.
   QW 125.5.S8 H799s 2006]   I. Title.
   QR82.S8H57 2006
   579.3'78—dc22   2006005669

9 8 7 6 5 4 3 2

Printed in the United States of America
on acid-free paper

To Joyce
for her companionship and encouragement
during more than four decades of marriage

Preface

Everyone has heard of antibiotics, and most people, at least in the developed world, have benefited from their curative powers. But how many of us know where they come from and how they developed into a cornerstone of medicine? The mold that famously contaminated Alexander Fleming's culture dish and eventually gave us penicillin is one of the icons of 20th century biology, but penicillin was just the first antibiotic to become a medicine. Dozens of important compounds followed, revolutionizing the treatment of infectious diseases. Most are made by a group of soil microbes, the actinomycetes, which were little known until their powers of antibiotic production were revealed, starting some 60 years ago.

This book begins by describing how these microbes were discovered and how they became an important source of antibiotics and moves on to an insider's account of how knowledge of their genetics developed over the second half of the 20th century. These insights, culminating in the determination of the complete DNA sequence for a model species at the start of the new millennium, have allowed us to understand the intricacies of actinomycete biology and the incredible feats of microengineering that go into building even a comparatively simple organism and adapting it superbly to its habitat. I describe how techniques for manipulating the genes for antibiotic production stemming from these studies are being applied to the challenge of making new antibiotics to counter the threat posed by pathogens that have become resistant to those in current use. Among these pathogens are other actinomycetes, relatives of the useful soil inhabitants, which cause deadly and disfiguring diseases: tuberculosis and leprosy. I talk about them too.

In attempting to bring the wonders of the actinomycetes to a wider audience I have tried to explain genetic concepts and fundamental biological principles in simple language, but I have included a glossary of terms for separate reference, and this may make some of the chapters intelligible in isolation.

I am indebted to the Leverhulme Trust for a grant to cover the costs of the project and to many people for their help and advice in writing this book. First and foremost my thanks go to my son, Nick Hopwood, who read two drafts and made innumerable suggestions for improving the manuscript. I should have been lost without his input. My wife, Joyce, made many valuable suggestions too, as did Jeffrey House of Oxford University Press. Douglas Eveleigh hosted a visit to the Waksman Archive and patiently answered my many subsequent questions about Rutgers University; Lisa Pontecorvo graciously gave me guided access to the archive of her father Guido; and Marianna Jackson devoted much time and effort to providing her reminiscences of life at Abbott during the Golden Age of antibiotic discovery. Many other colleagues generously responded to queries about specific topics: Boyd Woodruff for the early days of antibiotic discovery in Waksman's laboratory (Chapter 1); Liz Wellington for selective isolation of actinomycetes from soil, and Peter Hawkey for comments on clinically important antibiotic resistance (Chapter 2); Gilberto Corbellini for information on the Istituto Superiore di Sanità (Chapter 3); Natasha Lomovskaya for insights into science in Moscow before *perestroika* (Chapter 4); Stephen Bentley for many discussions about genome sequencing and the Sanger Institute (Chapter 5); Liz Wellington for spore dispersal, Geertje van Keulen for spore buoyancy, Jolanta Zakrzewska-Czerwinska and Dagmara Jakimowicz for chromosome replication and partition, and Carton Chen for chromosome transfer (Chapter 6); Marie-Joelle Virolle for amylase production, Hildgund Schrempf for chitin and cellulose degradation, Mark Buttner for vancomycin resistance, and Eriko Takano for signaling molecules (Chapter 7); Leonard Katz and David Cane for comments on Chapter 8; Cammy Kao for microarrays, Andy Hesketh for proteomics, and Kay Fowler for transposon mutagenesis (Chapter 9); and the late Jo Colston for answering my many questions about tuberculosis and leprosy (Chapter 10). I thank Keith Chater, Julian Davies, and Arny Demain for reading a draft of the whole manuscript and providing many useful suggestions.

I am greatly indebted to Tobias Kieser for generously providing many photographs and for teaching me the rudiments of Adobe Photoshop, and to Nigel Orme for imaginatively converting my rough sketches into the finished diagrams. I thank the many people, acknowledged in the captions, who provided other photographs. I am especially grateful to Helen Kieser for a long professional partnership, without which my own career would have been much less rewarding. I thank the many other colleagues at the John Innes Centre and worldwide who joined in the quest for knowledge about nature's antibiotic makers. Collaboration in science is nearly always beneficial, but in the *Streptomyces* field it has been unusually wide and prolonged, embracing commercial companies as well as universities and research institutes, and linking people across the world in a strikingly harmonious "family" that has helped to make my professional life both a happy and a satisfying one.

Contents

Introduction		3
1	Actinomycetes and Antibiotics	8
2	Antibiotic Discovery and Resistance	28
3	Microbial Sex	51
4	Toward Gene Cloning	81
5	From Chromosome Map to DNA Sequence	103
6	Bacteria That Develop	123
7	The Switch to Antibiotic Production	145
8	Unnatural Natural Products	165
9	Functional Genomics	193
10	Genomics Against Tuberculosis and Leprosy	211
Conclusion		226
Notes and References		229
Glossary		241
Index		245

STREPTOMYCES
in Nature and Medicine

Introduction

When we go for a walk in the woods, or spread compost on our garden, we smell a lovely earthy odor. If we visit the doctor with bronchitis or a septic finger, we will almost certainly be prescribed an antibiotic that will usually cure our symptoms in a few days, or if we are unlucky enough to get tuberculosis (TB) we will be put on a long course of antibiotic treatment. The connection between these topics is the subject of this book: microbes that live in the soil and make most of the antibiotics that are used around the world. These organisms, as well as evoking the outdoors, grow into strikingly beautiful colonies. They carry out amazingly complex processes on a tiny scale. They are among the most beautiful, fascinating, and useful of microbes.

The actinomycetes, as these microscopic chemists are called, manufacture antibiotics to help them compete with countless other microbes in the soil for space and food. By an amazing set of molecular switches, they sense the myriad opportunities and threats they meet in the soil and react appropriately. From the human perspective, antibiotic production is their most significant response. The actinomycetes were discovered in the last few decades of the 19th century, but they were a minority interest until streptomycin, the first really effective treatment for TB, was discovered in 1943. It was named after the most important genus of actinomycetes, *Streptomyces*.

Streptomycin was soon followed by a string of other *Streptomyces* antibiotics, establishing the actinomycetes as nature's chief antibiotic makers. Penicillin, the first antibiotic to be used in medicine, had been discovered by Alexander Fleming in 1928 and shown to be a life-saving drug during World War II. Compounds developed from penicillin are still crucially important in treating bacterial infections. Penicillin is made by a fungus, not an actinomycete, but actinomycetes make a much greater variety of antibiotics, including medicines to fight most bacterial and fungal diseases, as well as anticancer drugs and compounds that kill parasitic worms and insects. They and the fungi underpin an antibiotics industry valued at $25 billion a year today.

This book offers a personal view of the actinomycetes, based on a professional lifetime spent with them. As a PhD student in Cambridge in 1954, I began working on *Streptomyces*. My interest was, and still is, primarily a geneticist's, so genetics is the central theme of the book. Although the actinomycetes had been studied for decades as inhabitants of the soil, and in spite of the importance they were already showing as antibiotic producers, actinomycete genetics was still a virtual blank. I was attracted by the idea, prevalent ever since the first descriptions of the actinomycetes as a distinct group of microbes, that they might be missing links between the two major subdivisions of the natural world: the bacteria on the one hand and all the other organisms, including fungi, plants, and animals, on the other. I hoped that searching for genetic processes in the actinomycetes and comparing them with those in bacteria and fungi would throw light on the grand scheme of life.

When it started, microbial genetics was very low-tech, needing simple culture containers and minimal other equipment. Preeminent was the Petri dish, invented in the late 19th century and still the stock-in-trade of microbiology today. Originally of glass, now of disposable plastic, these shallow vessels are 9 cm (3.5 inches) in diameter, with a thin layer of jelly-like agar medium in the bottom and an overlapping lid. On the nutritive surface, a microscopic organism can soon multiply into millions of individuals, making a colony big enough to examine with the naked eye. Differences in colony shape or color are immediately obvious and can tell us about the inheritance of the genes that control them: I chose to work on a species called *Streptomyces coelicolor*, so called because it makes a beautiful blue pigment (*coelicolor* means "sky color" or "heavenly color" in Latin), because I hoped the pigment would provide a useful genetic handle, and it did. Other traits, biochemical, nutritional, and physiological, are not immediately detectable by inspection, but the geneticist can make them manifest. Microbes allow genetic studies to be performed on huge numbers of individuals in a small space, reproducing at a rate that corresponds in days to what would take weeks, months, or even years with a plant or animal. This is why microorganisms revealed much of what was learned about the molecular biology of genes during the second half of the 20th century, with new insights coming at an ever-increasing speed. It was an exciting time to be a geneticist.

Microbial genetics went through three phases during this period. In the first, the *in vivo* period, genes were studied in their native hosts, yielding a wealth of new knowledge about how the cell works. A simple bacterium called *Escherichia coli* that lives, usually harmlessly, in our intestines, became the main subject for this work, along with a few other microbes. The new knowledge was revolutionary. Whereas bacteria had been thought to reproduce only by simple, asexual splitting, they were found to exchange genes, but by processes very different from sex in higher organisms: as naked DNA, or through the agency of bacterial viruses, or in a bizarre kind of incomplete mating process. These natural gene exchanges were harnessed to great effect to discover genes and tell us a lot about how they work.

Then, in the mid-1970s, began an entirely new approach to genetics, recombinant DNA, which opened possibilities for understanding how organisms work at a level of detail previously unattainable. Scientists learned how to isolate a gene as pure DNA and analyze it down to its individual building blocks by determining their sequence along the molecule. In this *in vitro* phase, they figured out how to make new combinations of genes, or even artificial genes, in the test tube and introduce them into a microbe to study their effects, or to make useful medicines as the biotech industry was born.

Finally, in the mid-1990s, came the *in silico* phase of genetics, when the complete sequence of a microbe's DNA could be obtained in just a few months, later weeks or even days, and analyzed by computer. Gene functions could often be deduced from the sequence and confirmed by making changes in the DNA and following the outcome in a host microbe. *Streptomyces* genetics developed through all these phases, often using techniques and conclusions from other microbes as a guide and adding special tricks to reveal phenomena found in the actinomycetes but not in the simple *E. coli*.

The book is organized roughly historically. Chapter 1 relates how the first actinomycetes were described. They were mostly pathogens, starting with the leprosy bacillus in the 1870s and soon followed by the tubercle bacillus. It was some time before it was realized that they formed a natural grouping with soil-living organisms later called *Streptomyces*. In this chapter, I introduce Selman Waksman, a Russian immigrant to the United States in 1910, who developed a love for the actinomycetes in the 1920s and 1930s, when he led the rather unfashionable field of soil microbiology. He set the stage for the discovery of streptomycin as a cure for TB in the mid-1940s, catapulting the actinomycetes to a prominent place in industrial microbiology. In Chapter 2, I describe how, in the 1950s and 1960s, industrial scientists went about finding new antibiotics and bringing them to market during what became known as the Golden Age of antibiotic discovery. I discuss how the massive use of antibiotics, much of it of questionable wisdom at least in hindsight, led to a dangerous rise of antibiotic resistance that threatens the continuing efficacy of these wonder drugs.

With Chapter 3 we start to cover *Streptomyces* genetics from its origins in the mid-1950s. This chapter deals with its *in vivo* phase, during which I was one of a few lone researchers who discovered natural mating processes in *Streptomyces* that could be harnessed to map genes on the organisms' chromosomes and work out some of the mysteries of their genetics. Combined with advances in electron microscopy, these results revealed the true relationships of the actinomycetes. Rather than bridging the gulf between bacteria and higher organisms, they are true bacteria, but only distantly related to the others and with many special features. This chapter follows the early stages of my own career. After a decade in Cambridge, I spent 7 years at the University of Glasgow, Scotland, where I fell under the spell of Guido Pontecorvo, one of the luminaries of 20th-century genetics, who had promoted the idea of using microbial genetics to improve antibiotic productivity. Then, in 1968, I moved to Norwich to join an old-established institution, the John Innes Institute (later to be renamed the John Innes Centre), which had just relocated there to try to rejuvenate itself in a relationship with the University of East Anglia, one of the then new "plate-glass" English universities. This gave me the scope to build a research group big enough to delve deeper into the lives of the actinomycetes. Students, postdoctoral fellows, and visitors enlivened the laboratory, and I gained colleagues who went on to build research groups of their own.

During this period, as I relate in Chapter 4, we learned how to manipulate *Streptomyces* genes artificially, easing *Streptomyces* genetics into the *in vitro* phase. Meanwhile, the subject was taken up in universities and companies around the world, and a community of scientists grew as they shared their knowledge in publications and at meetings and summer schools. I experienced the excitement of being involved in this shared effort, with discoveries coming thick and fast. The skill and enthusiasm of this band of researchers moved *Streptomyces* genetics from a tiny minority interest on the fringes of the big stage of microbial genetics to a position nearer the limelight as many fascinating differences from the *E. coli* paradigm emerged.

With Chapter 5 we reach a turning point in the book and enter the *in silico* phase of *Streptomyces* genetics. Sequencing of the entire genome of *S. coelicolor* made it possible, at the dawn of the new millennium, to start analyzing its genetic endowment much more thoroughly. Chapters 6 and 7 illustrate how a complete genome sequence totally changes the way we think about the genetics, and indeed the whole biology, of any organism. I interpret some of the biology of *S. coelicolor* in light of the sequence: how it develops its characteristic form and physiology, including making antibiotics, to deal with the opportunities and threats it encounters. Although I showcase how *Streptomyces* adapts to its life in the soil, many of the principles of the story apply to bacteria that have evolved different life styles and adapted to different habitats, and indeed many of those principles originated in work on *E. coli*. Examples include the wonders of DNA replication, the molecular pumps that transport molecules in and out of the cells, and the relays of signals and responses that allow the organisms to switch on just those genes needed to deal with the situation of the moment. Since they arose before animals and plants and gave rise to them, bacteria are often regarded as primitive, but they have continued to evolve in the 4 billion years since they first emerged. Highly adapted machines containing minute systems of amazing complexity, bacteria are nature's nanotechnology.

Chapter 8 sees another shift of emphasis. Here I describe a new field of biotechnology that aims to use *Streptomyces* genetics to counter the threat posed by antibiotic resistance. Over recent decades, as it has become harder and harder to find effective antibiotics, the view has grown widespread that all the good natural compounds have already been discovered. Most of the big pharmaceutical companies therefore have turned back to their roots in synthetic chemistry and are trying to make entirely artificial drugs. But this approach does not seem to be working well for antibiotics. Using genetic engineering to make antibiotics related to but different from those found in nature looks to be a better strategy, and small biotech firms are eagerly exploiting the idea. Meanwhile, *Streptomyces* genomes are revealing innumerable genes for making antibiotics that were not anticipated. They encode the information for making compounds that the organisms manufacture only under special conditions that they may meet in the soil but are hard to reproduce in the laboratory. So a new challenge is to wake up these sleeping genes and find the compounds they make, which conventional antibiotic discovery campaigns overlook.

Even after extensive *in silico* analysis of any genome, we are left with thousands of genes with no assigned function, as well as many with tentative functions that need to be confirmed. Molecular biologists have started probing the roles of these genes by using a suite of techniques, collectively called functional genomics. With them we can learn how different sets of genes are switched on under varying conditions or at successive stages of the life cycle, thus gaining information about their real functions. This knowledge is complemented by investigating the consequences of systematically inactivating each gene. Chapter 9 shows how these techniques are being applied to *Streptomyces*. The results will throw a flood of new light on the biology of the organisms. At a practical level, this should help us to find the right conditions to express the sleeping antibiotic production genes and so discover useful new antibiotics that would otherwise have been missed.

In the final chapter, I return to two of the first actinomycetes to be discovered, the pathogens that cause TB and leprosy, and describe how genetics is helping to illuminate and combat the threat that these diseases still pose for mankind. As in *Streptomyces*, so in these pathogens: the genome is a mine of information, in this case about the strategies that the organisms use to avoid the defenses of the host and mount an attack that kills millions of people every year. Knowledge of these strategies will provide a major route to neutralizing them.

I hope that this book, about the first half-century of *Streptomyces* genetics, will give a feel for the excitement of being part of a community of scientists discovering the wonders of an amazing group of microbes. I hope it will introduce the accomplishments of these bacteria to a wider audience, including people with an amateur interest in science. Among professionals, perhaps those in the field will be interested in the history of some of the knowledge they use in their work. Other microbiologists will learn about a group of organisms they rarely meet. Hopefully, a wider group of biologists, as well as chemists, will also enjoy reading about the lives of these rarely seen but fragrant inhabitants of the soil and the ways in which they make the antibiotics that we usually encounter only as pills.

1

Actinomycetes and Antibiotics

At the start of the 20th century, medical bacteriology was a thriving study, after Louis Pasteur in France and Robert Koch in Germany had pioneered the germ theory of infectious disease in the preceding decades. Medical bacteriologists had clear objectives: to identify pathogenic bacteria and bring the diseases they caused under control. By comparison, soil bacteriology was a fragmented and unfashionable pursuit.

It was against this background that Selman Waksman began his career, investigating the biochemical capabilities of a previously obscure and poorly understood group of soil microbes and their contributions to agricultural fertility. As others had done before him, he grappled with the classification of this collection of apparently diverse organisms, described piecemeal from the 1870s onward. Halfway through Waksman's career came his discovery that these organisms, the actinomycetes, are nature's most prolific producers of antibiotics. One, streptomycin, was found to cure perhaps the most feared human disease, tuberculosis (TB), caused by the tubercle bacillus that Robert Koch had identified in 1882. As a result, the actinomycetes were to become some of the most important players in applied microbiology as the basis, along with the molds that make penicillin and related compounds, of a multibillion dollar antibiotics industry.

Selman Waksman and Soil Microbiology

Selman Abraham Waksman (1888–1973) was born in Novaia-Priluka, a small town in the Ukraine about 120 miles (190 km) from Kiev and 200 miles from Odessa. In his autobiography,[1] he gave a fascinating account of life in a Jewish family in rural, prerevolutionary Russia and how he developed an interest in books—first religious and later on secular subjects. He had a burning desire to learn and to better himself, but he had trouble overcoming the discrimination against Jews in taking official examinations and had no realistic chance of being accepted by the University of Odessa. Therefore, in common with countless others in a similar situation, he emigrated to the United States, staying first with a cousin in Philadelphia and then on a small farm belonging to another cousin and her husband, Mendel Kornblatt, in Metuchen, New Jersey. Figure 1.1 shows Waksman at the age of 21 in 1909, the year before he left home.

Waksman already had a broad interest in the chemical reactions that go on in living organisms and thought of taking a medical degree. He was accepted to study medicine by the College of Physicians and Surgeons at Columbia University in New York. Through helping on the farm, however, he was getting more and more fascinated by the science underpinning agriculture. Then Mendel Kornblatt suggested a meeting with Jacob G. Lipman, who was Head of the Department of Bacteriology at Rutgers College in New Brunswick, later to become the State University of New Jersey.

Waksman did not record whether Mendel knew Lipman personally, but Rutgers was only 10 miles from Metuchen, so perhaps they were acquainted as fellow members of the Russian immigrant community; or perhaps, as a farmer, Mendel simply knew Lipman by reputation. Rutgers was a major center for agricultural research and teaching. Affiliated to it was the New Jersey Agricultural Experiment Station, only the third such station to be set up in the United States, and Lipman was already an established figure. His leadership was acknowledged 10 years later by the founding of a dedicated College of Agriculture at Rutgers with Lipman at its head.

Figure 1.1. Selman Waksman in 1909. (Courtesy of Byron Waksman.)

Lipman's interests were primarily in soil bacteriology. In the early 20th century, research on soil bacteria had taken a very different path from the study of bacteria in the context of disease.[2] Doctors, veterinarians, and sanitary engineers worked with a limited range of pathogenic bacteria, mainly with a view to preventing them from causing disease or to curing the diseases if they occurred. In contrast, agricultural microbiologists were becoming more and more aware of the wide community of microbes in the soil and their potential to increase its fertility. Of particular concern was the availability of nitrogen. Two of the pioneers of soil microbiology, Sergei Nicolaevitch Winogradsky (1856–1953) in Russia and Martinus Willem Beijerinck (1851–1931) in Holland, had made their names in part with the discovery of bacteria that make nitrogen available to crop plants. Some convert ammonia, produced from the decomposition of proteins in plant and animal remains, into nitrite and then nitrate, the form in which nitrogen is normally taken up by plant roots. Others "fix" gaseous nitrogen from the air and make it available to plants, either symbiotically in nodules on the roots of leguminous plants such as peas, beans, and clover or as free-living inhabitants of the soil. Lipman, who had isolated some of the first free-living nitrogen-fixing bacteria, must have talked to Waksman enthusiastically about such microbial chemistry. It resonated with Waksman's interest in composting to increase soil fertility on the farm, and Waksman came away from the interview convinced that a degree in agriculture was the right thing for him. He won a scholarship to Rutgers and started there in 1912.

In his autobiography, Waksman was critical of some of his teachers and the courses he took at Rutgers—he credited Mendel Kornblatt on the farm with being his best teacher in his first year—but, as often happens at university, things looked up considerably in the final year, especially with his practical project. This consisted of taking soil samples from trenches dug on the Rutgers College farm and isolating microorganisms from them in the laboratory by spreading the samples on agar plates (Color Plate 1). Waksman wrote:

> At first, my main purpose was to count the bacterial colonies only. Here and there, however, there appeared also fungus colonies.... To my amazement, the agar plates also showed small colonies of organisms which were similar to those of the bacteria but under the microscope looked much like those of the fungi. These colonies appeared to be conical; they were leathery and compact when touched with a needle, and frequently pigmented when isolated and grown on different media. I immediately drew Dr. Lipman's attention to the great abundance of these colonies and asked for suggestions as to how to characterize and classify them. He confessed that he had never paid much attention to such colonies and considered them as some sort of bacteria, perhaps "higher bacteria." I appealed then to the plant pathologist, Dr. M. T. Cook, who taught me botany and mycology. We examined the colonies again and came to the conclusion that they represented an obscure group of little-known organisms, which usually were designated by the name *Actinomyces*. At the end of the year I tabulated my results. To my great amazement, these organisms showed a decided regularity of distribution in the soil. Their numbers depended entirely upon the nature of the soil, its reaction [acidity or alkalinity], depth from which the sample was taken, and the crop grown. Thus began my interest in a group of microbes [the actinomycetes] to which I was later to devote much of my time and which were to remain for the rest of my life my major scientific interest.

After graduating in 1915, Waksman decided to continue his study of "the soil and its life, especially my new-found friends, the actinomycetes." He became a research assistant in bacteriology at the New Jersey Agricultural Experiment Station and wrote his master's thesis in 1916, the year he became an American citizen. That year he also married Deborah Mitnik ("Boboli"), whom he had known in Russia. Then for 2 years he studied for a PhD at the University of California at Berkeley, just across the bay from San Francisco and one of the great centers of scientific research in the United States. His topic was enzymes from microorganisms, mostly actinomycetes. Toward the end of his stay in California, after the United States entered the First World War in 1917, he worked at a Berkeley company, Cutter Laboratories, helping to produce antitoxins and vaccines against bacterial infections.

In 1918, Waksman returned to New Jersey, where he was to spend the rest of his career. At first he had a 5-day-a-week job at the Takamine Company working on the synthetic chemical Salvarsan, the first effective treatment for syphilis, but his real interest was microbial life in the soil, and he spent 1 day of each week at Rutgers. He began to teach soil microbiology in Lipman's department, where he became associate professor in 1924, the year Rutgers College assumed university status. That year he went on a "grand scientific tour" to Europe, during which he met some of the pioneers of soil microbiology. Beijerinck greeted him with the words: "You are the Actinomyces man!"[1] He especially took to Winogradsky, who was born in Kiev, spent much of his career in St. Petersburg, and was now an émigré in Paris. They kept up a 30-year correspondence from that first meeting until Winogradsky's death. Waksman admired Winogradsky especially for emphasizing the importance of studying microbes in their natural habitat, where they existed, not in pure populations as they appear in the laboratory, but in complex communities of many interdependent species.

Waksman continued to do research on diverse aspects of life in the soil. In 1927 he published a 900-page tome on soil microbiology,[3] and in 1930 became full professor at Rutgers. During what he described as his "humus period," which lasted until 1939, he worked on the kinds of communities of microorganisms that Winogradsky had highlighted, elucidating their roles in breaking down plant and animal remains into humus. He published a long series of papers bringing together a mass of information about the effects of various kinds of soil microbes on each others' activities, thus establishing a formidable international reputation in soil microbiology. His real love, though, was "that obscure group of little-known organisms." What kind of organisms were they, and what were their relatives?

Discovery of the Actinomycetes

The name *Actinomyces* goes back to 1877, when it was applied to a microbe responsible for a disease of cattle called "lumpy jaw." It causes proliferation and distortion of the bone, resulting in incurable swellings on the side of the face that can eventually make it hard for the animal to eat (Figure 1.2). The disease used to be fairly common—I remember some spectacular lumps on the jaws of mature bulls when I worked on a farm as a boy—but is now rare, perhaps in part because cattle are usually

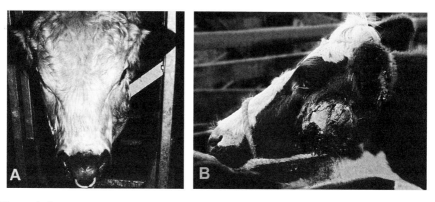

Figure 1.2. Cattle with lumpy jaw. (A) An early case in a British White bull. (B) An advanced case in a Holstein-Friesian cow. (Courtesy of Karin Mueller, University of Cambridge.)

not kept as long as they used to be. A German botanist, Carl Otto Harz (1842–1906), working at the Royal Veterinary School in Munich, first described the causal agent in a lecture in May 1877 and wrote about it in the yearbook of the school.[4] The bony lesions are roughly spherical and develop radial striations as they increase in size with growth of the organism. Long, thin filaments are visible in the center, while the outer layers show more regular ray-like structures that seem to end in club-shaped bodies (Figure 1.3A). Harz interpreted these as "gonidia," typical of the reproductive bodies of certain fungi, and the structures bearing them as fungal hyphae (Figure 1.3B). He therefore described the microorganism as a fungus and called it *Actinomyces bovis* (*Actinomyces* means "ray fungus"). Harz had to confine his study to a morphological description of what he saw in the animal tissues and could not obtain a pure cul-

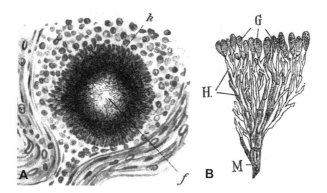

Figure 1.3. (A) *Actinomyces bovis*, seen in sections through nodules in a cow's jawbone. *f*, mycelial filaments; *k*, ring of swollen cells. (From Lehmann, K. and Neumann, R. [1896]. *Atlas und Grundriss der Bakteriologie* [Atlas and outline of bacteriology]. Munich: J. F. Lehmann.) (B) Detail of the outer layers of a nodule. G, "gonidia"; H, hyphae; M, mycelial cell. (From Harz, C. O. [1877–1878]. *Actinomyces bovis,* ein neuer Schimmel in den Geweben des Rindes. *Jahresbericht der Kaiserlichen Central-Thierarznei-Schule in München,* 125–140.)

ture (probably, with hindsight, because the organism grows in the absence of oxygen and dies on prolonged contact with air). This was unfortunate, because otherwise he would doubtless have realized that the fine filaments seen in the center of the lesions represent the microorganism, while the obvious structures on the outside are actually host cells.

Harz was not the first to discover an organism that would eventually become known as an actinomycete. This happened in Norway just before Harz described his "ray fungus." Leprosy was common in Europe in the Middle Ages and then mysteriously declined. Surprisingly, because we now think of leprosy as a tropical disease, its final European hideout was in one of the coldest countries. The last recorded leprosy patient was an old man who was admitted to the hospital in Bergen from one of the small islands off the coast in 1962 or 1963 with gangrene of a toe.[5] In the 19th century, leprosy was rife among the poor in the region around Bergen, which became a center of attempts to understand the disease. Even though lepers had been ostracized in Europe for centuries, at least in part to prevent others from catching the disease, a popular view was that leprosy was inherited. Armauer Hansen (1841–1912), who joined the staff of the Bergen Leprosy Hospitals in 1868 soon after completing his medical training, was the first person to identify the causal agent as a microorganism. In a long article in 1874,[6] he described his discovery of the leprosy organism, which is still called "Hansen's bacillus" today. Figure 1.4A shows a later drawing of the organism, a tiny rod-shaped microbe with a slightly wavy outline and rather irregularly shaped cells.

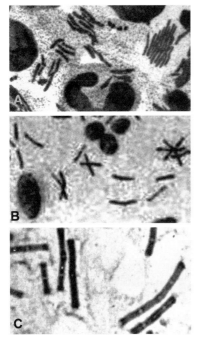

Figure 1.4. (A) *Mycobacterium leprae.* (Drawing from Lehmann, K. and Neumann, R. [1896]. *Atlas und Grundriss der Bakteriologie.* Munich: J. F. Lehmann.) (B) *Mycobacterium tuberculosis.* (Drawing from Koch, R. [1884]. Die Aetiologie der Tuberkulose. *Mitteilungen aus dem Kaiserlichem Gesundheitsamte* 2, 1–88. (C) *Bacillus anthracis.* (Photomicrograph from Koch, R. [1877]. Verfahren zur Untersuchung, zum Conservieren und Photographieren der Bakterien. *Beiträge zur Biologie der Pflanzen* 2, 399–434.) The large, rounded bodies in A and B are nuclei of host cells.

Hansen could not prove that he had identified the real cause of leprosy, rather than an organism associated with disease symptoms caused by something else. To convince his colleagues, he had to rely on indirect evidence, much of it gleaned from a National Registry for Leprosy that had been compiled for Norway in 1856. Hansen analyzed the incidence of leprosy in extended families, where he found no clear pattern of inheritance. Then there was a mass of data showing that leprosy was much more widespread in the countryside than in the towns. Hansen pointed out that rural communities often used no bed sheets, and different people typically occupied the same bed in succession, whereas in the towns more people had their own beds and used laundered sheets. Finally, when Hansen probed the memories of leper patients going back as long as 7 years before they presented with the disease, they almost always recalled contact with a leprosy sufferer.

Perhaps not surprisingly, most of Hansen's colleagues, including the director of the hospital, Daniel Danielssen, were not convinced and stuck to the idea that leprosy was inherited, although this did not apparently cause a rift between the two men: Hansen married one of Danielssen's daughters in 1873. Hansen tried inoculating his bacillus into rabbits, but they showed no disease symptoms. He became increasingly frustrated and in 1879 attempted to infect the eye of a woman who was already suffering from neural leprosy with material from another patent with leprosy of the skin. He had published a monograph on leprosy of the eye with an Oslo ophthalmologist, O. B. Bull, in 1873[7] and evidently had a special interest in this form of the disease. But he failed to obtain the patient's consent or explain why he was doing it, so the city authorities took him to court on behalf of the patient and won a claim for damages. He lost his job as resident physician at the Bergen Leprosy Hospitals, and, although he remained Medical Officer for Leprosy for the whole of Norway, that was the end of his research career.

One of the pioneers of 19th century bacteriology was Ferdinand Cohn (1828–1898)[8] (Figure 1.5), who established an Institute of Plant Physiology in Breslau in what was then German Silesia; it has since reverted to its Polish name of Wrocław. In 1875, Cohn published a treatise summarizing his observations on a whole range of microbes, including one he called *Streptothrix foersteri* after a medical friend, R. Foerster, who had supplied him with the material from infected human tear ducts in which he saw the organism.[9] This microbe was not clearly implicated in any disease, however, and with hindsight it had probably been blown into the patient's eye on a soil particle. It had a much more complicated structure than Hansen's bacillus, with elongated, branching cells reminiscent of those of fungi, but on a minute scale (Figure 1.6). The organism was present along with various typical spherical bacteria, and Cohn could not separate it from them and grow it in pure culture. Nevertheless, his account of the organism is a milestone in the history of microbiology, because it is now recognized as the first description of a soil-living actinomycete of the kind that Waksman would later spend his career studying.

Robert Koch (1843–1910)[10] was the next to describe what we now consider as an actinomycete, the tubercle bacillus, but this was not his first contribution to the young science of bacteriology. He had a medical practice in Wollstein (Wolstyn), now in Poland but part of Germany at the time, where he was also District Medical Officer, but he began to do research as a hobby. In 1876, he published a seminal paper prov-

Figure 1.5. Ferdinand Cohn, who described the first organism that we would now call a *Streptomyces*. (Courtesy of Gerhart Drews, University of Freiburg; photograph originally published in Cohn, P. [1901]. *Ferdinand Cohn, Blätter der Erinnerung*. Breslau: J. U. Kern.)

ing, perhaps for the first time, that a microbe was the cause of a disease. This was anthrax, which mainly infects farm animals. The anthrax bacillus is a typical bacterium, with regular rod-shaped cells (Figure 1.4C), not an actinomycete. The bacillus had already been discovered, but it was Koch who showed that it alone could cause anthrax. However, he felt the need for reassurance from an established scientist that he was on the right lines, so he visited Cohn, as a respected senior bacteriologist, to validate his ground-breaking conclusions on anthrax before he committed himself to them in public (Wolstyn is only 80 miles from Wrocław, so it was an easy trip). Cohn was happy to lend support to Koch's results—in fact, he was most impressed and excited by them—and provided crucial encouragement to Koch in his next major endeavor, to find the cause of TB.

Koch turned his attention to TB in 1881, after moving to the Imperial Health Office in Berlin. In an amazingly short time, he identified the tubercle bacillus. Like leprosy and many other diseases, TB had been attributed to all kinds of causes, but in 1865 a French physician, Jean-Antoine Villemin, had shown that the disease could be transmitted to experimental animals from the tissues of patients. Koch isolated and cultivated the culprit and inoculated it into guinea pigs, where it caused TB. He announced his results on March 24, 1882, in a lecture at the Berlin Physiological Society and in a short paper in print only 3 weeks later.[11] To describe the tubercle bacillus unambiguously so that the world would accept his findings, Koch had to stain the organisms to allow clear drawings to be made. He took advantage of artificial dyes that the German chemical industry was developing for textiles. However, he needed to use a caustic solution to allow the dye to penetrate the cells. This was quite different from, and much more difficult than, the staining of bacteria Koch had worked

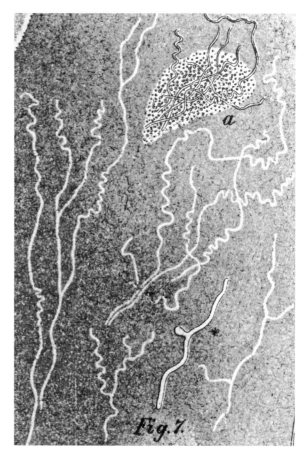

Figure 1.6. Drawings of *Streptothrix foersteri*. The image labeled "*a*" shows the thread-like *Streptothrix* filaments embedded in a mass of unicellular bacteria ("micrococci"); these bacteria were washed away to give the other images; the asterisk identifies "a thicker thread resembling mycelium." (From Cohn, F. [1875]. Untersuchungen über Bakterien. II. *Beiträge zur Biologie der Pflanzen* 1, 141–204.)

with before, such as the anthrax organism, so he deduced that the tubercle bacillus must be protected by an outer layer with unusual properties. The microbes were thinner and less regular in outline than the anthrax bacillus, and often curved, sometimes "to such an extent as to reach the first stage of a corkscrew structure." Koch likened them to those that Hansen had proposed to cause leprosy: as Figure 1.4B shows, they look remarkably similar to Hansen's bacillus (Figure 1.4A).

In the following decades, many other microbes were described, notably the soil inhabitants that give rise to the "leathery and compact" colonies that had puzzled Waksman in his early work at Rutgers. The colonies have these characteristics because they consist of interconnected branching cells like those that Cohn had described as *Streptothrix*. The elongated cells are called hyphae, and the mass they form is a mycelium. These same terms are used for the corresponding, but larger, components of fungal colonies. In contrast, colonies of the tubercle bacillus are soft and pasty because they consist largely of individual cells. Yet there seemed to be a relationship among the various organisms.

The group was recognized officially with the Latin name *Actinomycetales* in 1916 by R. E. Buchanan,[12] a bacteriologist at Iowa State College, but this did not put an end to arguments about the classification of the group and its boundaries. Waksman contributed to the discussion from 1919 onward. However, none of the schemes caught on. Arthur Henrici, a professor of bacteriology at the University of Minnesota, made a telling comment in 1930[13]: "It is difficult to read the literature [on the actinomycetes] intelligently because of the multiplicity of names which have been applied, sometimes to the group as a whole, sometimes to portions of it." Classification of a new group of organisms is often confusing at the start, but the actinomycetes were an extreme case.

In 1943, Waksman and Henrici tried again to classify the actinomycetes.[14] They proposed that the capacity to form branching cells was the hallmark of the actinomycetes and used the degree of branching to define three major groups, two of which they subdivided, giving five genera altogether (Figure 1.7). The first contained the leprosy and tubercle bacilli, which mostly grew as single rod-like cells but from time

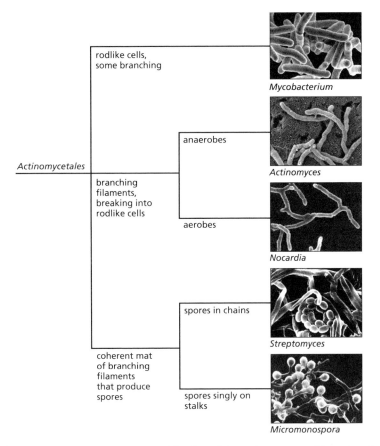

Figure 1.7. Waksman and Henrici's classification of the actinomycetes. (Scanning electron micrographs were kindly supplied as follows: *Mycobacterium*, Clifton Barry, National Institutes of Health, Bethesda MD; *Actinomyces* and *Nocardia*, Yuzuru Mikami, Chiba University, Japan; *Streptomyces*, Kim Findlay, John Innes Centre; *Micromonospora*, Yasuhiro Gyobu, Meiji Seika Company, Japan.)

to time formed side branches; the name *Mycobacterium*, given to the two pathogens by Karl Lehmann and Robert Neumann in 1896,[15] was retained for them. Organisms that developed a mass of branching filaments, which later broke up into individual cells, were classified in a second group. Some of these, like the lumpy-jaw organism, grew in the absence of oxygen: they were anaerobes. This was deemed to be an important characteristic, so these organisms were placed in a separate genus and the name *Actinomyces* was retained for them, while organisms that required oxygen—aerobes—were called *Nocardia*. This name had been used for various microbes since its introduction to honor a French microbiologist, Edmond Nocard, who had described the causal agent of a disease of cattle in Guadeloupe. The third and final group consisted of microbes that developed a dense mat of interconnected branching hyphae that remained intact and gave rise to specialized reproductive spores. Two genera were proposed for them, depending on whether the spores were produced in chains or singly, on stalks sprouting from the sides of the main hyphae. Cohn's *Streptothrix* fell into the former group, but it turned out that his use of the name was invalid because it had been applied earlier to a totally different microbe, so Waksman and Henrici invented a new one, *Streptomyces*. A previously used name, *Micromonospora* (meaning "small, single spores"), was adopted for the second group.

The purpose of classifying organisms is to reflect how they fit into the natural world. Waksman and Henrici had come up with a simple, pragmatic classification of the actinomycetes, but it did not address their relationships to other organisms. Were the actinomycetes bacteria that had evolved a more complicated growth form than the simple rod-shaped or spherical cells typical of bacteria? Or were they fungi that had cellular dimensions much smaller than a typical fungus? Or were they in fact a truly intermediate group? The names given to many of the organisms reflected this ambiguity. The word *Mycobacterium* means "fungus bacterium," because their cells looked like bacteria but had the wavy outline and capacity to branch characteristic of fungi. Harz had called his organism "ray fungus," because the cells looked like fungal filaments radiating from the middle of each lesion, and he erroneously thought that they gave rise to fungal "gonidia." *Streptomyces* means "twisted fungus," replacing Cohn's name *Streptothrix* ("twisted hair"). Waksman tended to favor the idea of the actinomycetes as a group intermediate between bacteria and fungi, but, as he was well aware, the information needed to decide the question did not exist at the time. In any case, the huge evolutionary gulf that separates bacteria from fungi and higher forms of life, including plants and animals, had not yet been appreciated, so the issue was not a burning one. It was not resolved until much later (see Chapter 3).

Even though Waksman could not be certain of the relationships of the actinomycetes to other organisms, or even of the precise boundaries of the group, it was mainly because of his passionate interest in them as key members of the community of microorganisms in the soil that knowledge of them advanced during the 1920s and 1930s. Nevertheless, they remained rather a Cinderella group until, with the discovery of streptomycin, the world suddenly had to take them seriously. Waksman had not been directly interested in disease-causing microbes until just before the start of the Second World War, although probably his stints at the Cutter and Takamine companies had left some mark. Suddenly he switched the efforts of his laboratory at Rutgers (Figure 1.8) to a hunt for compounds made by soil microorganisms that would kill bacteria causing human disease.

Figure 1.8. The building at Rutgers where Waksman had his laboratories, around 1926: the building in the middle distance, now called Martin Hall. On the right is the Administration Building. (Courtesy of Waksman Archive.)

The Discovery of Streptomycin

Waksman's decision to look for antimicrobial agents stemmed directly from two events. Alexander Fleming discovered penicillin in 1928, but he could not develop it as a means of killing bacterial infections: it seemed to be very unstable, and in any case Fleming was not a chemist. It took 10 years before Howard Florey's group in Oxford was able to isolate the compound in pure form from cultures of the fungus that made it. Ernst Chain began to take an interest in it in 1938, and by 1941 he and other members of the Oxford team had shown it to be a life-saving antibacterial drug.

René Dubos made the other momentous discovery. He had qualified in agriculture in Paris but was so impressed on hearing Winogradsky speak in 1924 about soil microbiology at a conference in Rome, where Dubos was working as an editor at the International Institute of Agriculture, that he switched fields. He had met Waksman at the conference and got to know him on the boat that they both took to the United States, Waksman returning to Rutgers and Dubos starting a new life in America. He became a PhD student with Waksman, working on the breakdown of cellulose by soil microbes. Dubos then moved to the Rockefeller Institute in New York where his first project, with Oswald Avery, was a deliberate search for an enzyme to destroy the protective coat of the bacterium that causes pneumonia. Later, he set out to find an agent produced by a soil microbe that would kill other pathogenic bacteria. He added the pathogens to soil in the hope that the microorganisms there would attack them and multiply by feeding on the dead cells, and in 1939 he found an antibacterial agent that he called tyrothricin, made by a culture of *Bacillus brevis*.[16] Whereas Fleming discovered penicillin when a fungus happened to contaminate one of his culture plates, tyrothricin was the fruit of a deliberate search by Dubos for an antibacterial agent.

In collaboration with the chemist Rollin Hotchkiss at the Rockefeller Institute, Dubos (Figure 1.9) showed tyrothricin to be a mixture of two compounds, tyrocidine and gramicidin. The first was immediately found to kill mice as well as bacteria, but gramicidin appeared much more promising. Its discovery caused a great stir when it was found to cure experimental infections in animals, but it too turned out to be too toxic for use in medicine, except for surface application to infected tissues: it could not be swallowed or injected. In any case, its promise was soon eclipsed by penicillin, which was amazingly safe, but the Oxford group recognized Dubos's discovery as a prelude to their own work.

The excitement engendered by gramicidin and then penicillin had a profound effect on Waksman, who "became fully convinced that all my prior knowledge of the fungi and actinomycetes, of their occurrence and activities, gave me just the tools required for this type of research." He was spurred on by his conviction that different kinds of microbes compete with each other in the soil and that they might have evolved

Figure 1.9. René Dubos speaking at the official opening of the Rutgers Institute of Microbiology (later the Waksman Institute) with Selman Waksman, 1954. (Courtesy of Waksman Archive.)

chemical agents to help them gain an advantage. This idea was reinforced by the observation that disease-causing bacteria survive for only short periods in the soil, suggesting that the soil microbes might actively destroy them. The soil should therefore be the best place to look for antibacterial agents. The term "antibiotic" had been coined in 1889 by a French biologist, P. Vuillemin, but applied very generally to describe the destruction of one organism, not necessarily a microbe, by another. Waksman suggested using the word very precisely to distinguish naturally produced inhibitors from the synthetic compounds that had long been used, with mixed success, to combat infectious agents, either as general disinfectants such as phenol or as therapeutic compounds such as Salvarsan or the then recently discovered sulphonamides. Waksman's definition of antibiotics was, "chemical substances that are produced by microorganisms and that have the capacity, in dilute solution, to selectively inhibit the growth of or even to destroy other microorganisms."[17] The phrase "in dilute solution" was added to the original definition to exclude substances such as ethyl alcohol, which is produced by yeasts and kills other microbes, but usually only at a relatively high concentration, such as a 10 percent solution.

Penicillin, although truly a "wonder drug," killed only certain kinds of bacteria. Bacteria are classified into two major groups, gram-positive and gram-negative, depending on their ability to be stained for microscopy by a dye system invented by a Danish bacteriologist, Hans Christian Gram, in 1884. Although invaluable for diagnostic purposes, there is a lot more to this apparently trivial distinction. It reflects a fundamental difference in the architecture of the cells. In all forms of life, the cell contents are surrounded by a membrane, which lets appropriate kinds of molecules in or out—nutrients in, waste products out, for example—and thereby maintains the integrity of the cell. In bacteria, the cell membrane is surrounded in turn by a rigid wall, which gives the cell its characteristic shape—spherical, rod-shaped, or filamentous. Gram-negative, but not gram-positive, bacteria have an extra membrane, outside the wall, and this makes them harder to kill with antibiotics, which often cannot cross the outer membrane.

Penicillin proved very effective against gram-positive pathogens, such as the *Staphylococcus* that causes septicemia (blood poisoning) or *Streptococcus* species that give rise to scarlet and rheumatic fevers, but not against gram-negative bacteria such as the *Salmonella* that causes typhoid fever, the *Vibrio cholerae* responsible for cholera, or bacteria that cause urinary tract infections. The outer layers of the tubercle bacillus are also extremely hard to penetrate, as Koch had found when he described special conditions for staining the organism. Therefore, when Waksman turned his attention to finding new antibiotics, he disregarded gram-positive bacteria as targets—penicillin was taking care of them—and sought an effective agent to treat gram-negative infections, with TB as a second goal.

At first Waksman's group tested all three groups of soil microbes—bacteria, fungi, and actinomycetes—for their potential to kill the target germs. They inoculated each candidate antibiotic producer across a Petri dish of nutrient medium and streaked the pathogens at right angles to it; antibiotic production was indicated when growth of the pathogens was inhibited by material diffusing from the producing organism (Figure 1.10). It was soon apparent that the actinomycetes were the most productive group, and amazingly so. In one study, of 244 actinomycete cultures isolated at ran-

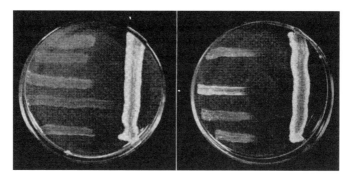

Figure 1.10. The cross-streak method to detect antibiotics. The actinomycete to be evaluated for antibiotic production has been streaked on the right hand side of each Petri dish, with bacteria to be tested for sensitivity at right angles. Note inhibition of these bacteria to varying degrees by antibiotic diffusing from the actinomycete growth. (From Waksman, S. A. [1950]. *The Actinomycetes: their nature, occurrence and importance*. Waltham MA: Chronica Botanica Co.)

dom from different soils, 106 showed some antibiotic properties in such tests, and 49 were "highly antagonistic." One actinomycete gave them their first promising lead, in 1940. It made a red material that they named actinomycin after Harz's lumpy jaw organism, *Actinomyces*. The graduate student H. Boyd Woodruff isolated the producer from a pot of soil to which he had added bacteria over a period of 3 months in the hope that this would favor actinomycetes that could kill and feed on the bacteria, just as Dubos had done when he found tyrothricin. Whether or not this helped in the discovery of actinomycin will never be known, because there was no control experiment without the added bacteria.

Waksman's group now needed to make enough actinomycin to study its properties. Waksman wrote in his autobiography (perhaps in part with the benefit of hindsight):

> Accompanied by two of my assistants, I visited Merck & Co. . . . When we isolated actinomycin, I wanted to utilize the large tray facilities of that company for growing microorganisms, to produce a large amount of the antibiotic required for chemical and animal studies. When we were about to inoculate the large chambers with a spore suspension of the actinomycin-producing organism, I said to my assistants: "You are now witnessing an historical event. It is the first attempt that has ever been made to grow an actinomyces on a large scale, to attempt to utilize an actinomyces for any practical purpose, or even to find any use whatsoever for this obscure group of micro-organisms."

Unfortunately, the initial excitement did not last long: actinomycin certainly killed the pathogens, but it killed laboratory animals too. Woodruff pressed on and isolated the next promising antibiotic candidate, also from an actinomycete, the following year. Named streptothricin after Cohn's *Streptothrix*, it cured experimental infections in mice. An improvised clinical trial in humans began but was interrupted straightaway because streptothricin was found to cause kidney damage—Woodruff recalled that the four treated subjects stopped urinating within a few hours; fortunately, they recovered when the treatment was stopped.[18] Streptothricin was soon found to cause delayed toxicity in mice.

Then, in 1943, came the breakthrough. Another graduate student, Albert Schatz, found an antibiotic produced by a strain of *Streptomyces griseus*, the species that Waksman and Henrici had chosen as the type member of their new genus that same year. Streptomycin killed gram-negative pathogens, but soon Schatz made the momentous discovery that it was also effective against the tubercle bacillus. Larger quantities of the antibiotic were urgently needed, and a mushroom farm was rented and adapted to produce streptomycin in the room that had been used to inoculate the mushroom containers[19]—a far cry from the rigorous conditions for drug making now demanded by the Food and Drug Administration in the United States and similar regulatory bodies in other countries to ensure that medicines are produced as hygienically as possible.

The first announcement of streptomycin in a scientific paper was in January 1944. Later that year, two experts in TB, William Feldman and Corwin Hinshaw at the Mayo Clinic in Minnesota, began to evaluate it as a treatment for TB in guinea pigs, which are very susceptible to human strains of the pathogen. By the end of the year they had shown streptomycin to be extremely promising, and a detailed report of the guinea pig studies appeared in 1945 (Figure 1.11). Trials on patients at the Mayo Clinic and elsewhere progressed rapidly, mostly using streptomycin produced by Merck (Figure 1.12), and by the end of 1946 there was a report by the Committee on Chemotherapy of the U.S. National Research Council about the first 1000 TB patients treated. This and other results showed that the new drug really cured cases of the disease.

Merck had rights to exclusive licenses for work in Waksman's laboratory, which they had supported since 1938, when Waksman helped them develop a fungal fermentation process to produce citric acid. They were persuaded that this arrangement was not appropriate for a life-saving drug that suddenly was needed in large quantities, and soon eight U.S. companies had taken licenses. By mid-1947, they were producing 1000 kg of streptomycin a month. The all-time high for the annual value of streptomycin produced in the United States (and by then other countries were also making large amounts) was 47 million dollars in 1951,[20] a large sum in those days. Royalties were paid to a new body, the Rutgers Research and Education Foundation, and by 1954 the Foundation had built a new Institute of Microbiology for Waksman (Figure 1.13).

Streptomycin and its power to cure TB led to fame for Waksman, marred by controversy when Schatz complained that he had been denied a fair share of acclaim and financial reward for the discovery. The resulting litigation led to a pretrial settlement in 1950, establishing Schatz as the legal and scientific codiscoverer of streptomycin and giving him 3% of the royalties (Waksman received another 17%, of which he distributed 7% to other members of his laboratory). Waksman was awarded the 1952 Nobel Prize for Physiology or Medicine. Although few regarded this as undeserved, many expressed the view that it should have been shared, with Schatz, or with Jorgen Lehmann for his work in the early 1940s in Stockholm that led to the development of a synthetic derivative of aspirin, para-aminosalicylic acid (PAS), to treat TB. Its utility was overshadowed by streptomycin, but a combination of PAS and streptomycin turned out to be a more powerful treatment for TB than streptomycin alone.[21]

Over the decades, the controversy over Schatz's contribution to the discovery of streptomycin became a *cause célèbre*, with protagonists on both sides of the argument.[22]

CONTROLS

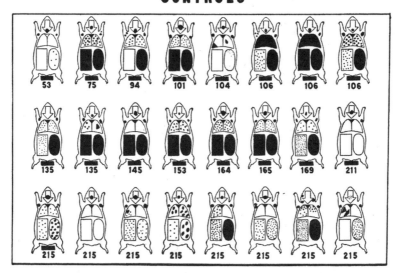

STREPTOMYCIN SERIES
TREATED AFTER 49 DAYS

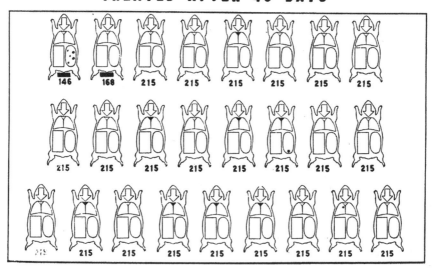

Figure 1.11. The seminal experiment showing curing of tuberculosis in guinea pigs by streptomycin. The organs are represented schematically as different shapes, with lungs at the top, liver lower left, and spleen lower right. Black shading represents intensity of infection. A black dot on the arrow indicates a lesion at the site of infection, and dots at the base of the forelimbs indicate lymph node involvement. Numbers below black bars are days at death after inoculation with *Mycobacterium tuberculosis*; other animals survived to autopsy at 215 days. (From Feldman, W. H., Hinshaw, H. C. and Mann, F. C. [1945]. Streptomycin in experimental tuberculosis. *American Review of Tuberculosis* 52, 269–298.)

Figure 1.12. Waksman at the building of Merck's first large-scale streptomycin plant, in Elkton, Virginia, 1946. The person on the right is believed to be Dr. Russel Aikens of Merck. (Courtesy of Waksman Archive.)

Figure 1.13. The Waksman Institute at Rutgers, January 2003. (Courtesy of Douglas Eveleigh, Rutgers University.)

Was Schatz, however skilled, simply the person who happened to study the crucial *S. griseus* cultures, out of all those being screened in the laboratory that Waksman had set up with the specific aim of finding novel antibiotics? Or did he make a truly inventive contribution to the discovery, working essentially on his own in a basement laboratory that Waksman hardly ever visited? How can we decide? Nobel Prizes are usually judged to be well deserved, but they honor only a small fraction of potential recipients, and those who (narrowly?) miss out on them can be deeply affected, as Albert Schatz undoubtedly was, right up to his death in February 2005 at the age of 84, even though he was recognized by Rutgers University with its highest honor, the Rutgers University Medal, in 1994. Many believe that René Dubos was unlucky not to be a Nobel laureate, but he does not appear to have borne a grudge.

The discovery that streptomycin could vanquish the tubercle bacillus led to excited press reports. For example, an article in the *New York World-Telegram* of February 3, 1947, was headed, "Mold drug new weapon in fight on tuberculosis." But on February 19 the same newspaper ran a column entitled, "Is streptomycin the atom bomb in TB war? Scientists study why it has proved a dud in some tuberculosis cases." Evidently, the drug was not going to banish the disease forever, even if it saved the lives of many

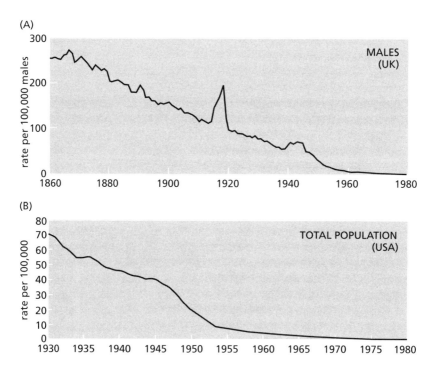

Figure 1.14. Mortality caused by tuberculosis. (A) Mortality among males in the United Kingdom, 1860–1980. (Redrawn from Vynnycky, E. and Fine, P. E. M. [1998]. The long term dynamics of tuberculosis and other diseases with long serial intervals: implications of and for changing reproduction numbers. *Epidemiology and Infection* 121, 309–324.) (B) Mortality among the total population in the United States, 1930–1980. (Redrawn using data from a web page of Joseph M. Mylotte, State University of Buffalo, NY.)

individual sufferers from an illness that had been a scourge of mankind for so long. The tubercle bacillus was already becoming resistant to streptomycin.

I have described in this chapter how various microorganisms identified in the last quarter of the 19th century came to be recognized as a special group, which was given the name actinomycetes. Their relationship to other organisms was unclear, and their study was not a mainstream topic of biological research, or even of microbiology. More or less by chance, Selman Waksman chose them as his main subject in the context of the relatively unglamorous science of soil microbiology. For 25 years, he maintained an interest in them before they suddenly moved to center stage with the discovery of the first really effective treatment for TB. It is a nice coincidence—though no more— that the producer of streptomycin, *S. griseus*, and the deadly pathogen that it was found to kill, *Mycobacterium tuberculosis*, are both members of the actinomycetes.

Streptomycin did not eradicate TB. In developed countries, the disease had been in more or less continuous decline for more than a century (with temporary increases in Europe during the two world wars), and against this lowered incidence, raw mortality figures showed only a small dip after streptomycin was introduced in 1947 (Figure 1.14), although many individuals owed their lives to its power. However, the discovery of streptomycin had a much more far-reaching consequence. It showed that the first clinically important antibiotic, penicillin, was not a flash in the pan and that other useful antibacterial compounds awaited discovery. Soon, a huge effort was underway to find more medical marvels from the previously obscure actinomycetes. It was hoped that some would overcome the resistance of the TB pathogen to streptomycin and that others would kill different groups of disease-causing germs. A new industry developed to cash in on this potential, as described in the next chapter.

2

Antibiotic Discovery and Resistance

The discovery of three potent actinomycete antibiotics in as many years made these microbes a happy hunting ground for further valuable drugs to fight human infections. Only streptomycin was nontoxic enough for medical use, but odds of one in three was still a good bet. After streptomycin's success, Waksman's group therefore continued isolating and screening actinomycetes, finding a dozen more antibiotics over the next 10 years. They were not alone.

Pharmaceutical companies had taken up the challenge of developing efficient methods for making penicillin and streptomycin in the 1940s, but they were not at first looking for new antibiotics themselves. All this changed in the latter part of the decade, as almost all the big American companies started their own screening programs. So did companies in other countries, as well as academic groups, especially in Japan. Through all this activity, many new antibiotics were identified in the 20 years from the late 1940s to the late 1960s, mostly from the actinomycetes. Enough of the compounds became successful drugs that the industry burgeoned. The rate of antibiotic discovery then declined sharply, so these productive years came to be called the Golden Age (Figure 2.1). Its legacy was a revolution in the treatment of infectious disease. Most bacterial pathogens were brought under control, but mushrooming antibiotic use also led to a dramatic rise in antibiotic resistance as the disease-causing bacteria fought back. It became clear that antibiotic discovery was not going to be a once-and-for-all activity to vanquish the pathogens, but an ongoing quest to find new treatments for old infections.

In this chapter, I describe what makes a good antibiotic and how companies went about finding them and bringing them to market during the Golden Age. I then talk about how antibiotic resistance arises, why it is prevalent, and how to minimize it.

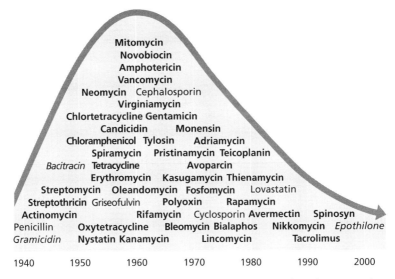

Figure 2.1. Discovery of important antibiotics and other natural products over the years. **Bold type** indicates actinomycete products; normal type indicates fungal products; *italic type* indicates products from non-actinomycete bacteria.

The chapter ends with an introduction to approaches for developing new antibiotics, mainly to combat the rise of acquired resistance.

How Do Antibiotics Work?

Antibiotics act by inhibiting specific cellular targets, and their ability to kill pathogens without harming the host depends on the nature of the target. For penicillin it is the bacterial cell wall, composed of long chains of sugar molecules cross-linked by short bridges made of amino acids, the same substances that, in much longer chains, build proteins. Penicillin blocks the enzymes that build the bridges, so the cell wall is fatally weakened; it ruptures as the bacterium grows, literally bursting at the seams and spewing out the cell contents. Because there is no counterpart to the bacterial cell wall in animals, penicillin does not affect them. Streptomycin might be expected to be toxic because it inhibits protein synthesis, a process common to all forms of life. Proteins are made on the tiny cell factories called ribosomes, which receive the sequence of letters of the genetic code, carried to them from the DNA of the chromosomes in the form of an RNA message, and translate them into the correct sequence of amino acids to make a specific protein (Figure 2.2). Streptomycin blocks translation by binding to bacterial ribosomes; the subtly different human ribosomes have no affinity for streptomycin, so it is not toxic. Actinomycin blocks a different step in gene expression: it binds to the DNA, preventing access of the transcribing enzyme, RNA polymerase. Because any DNA is subject to such binding, actinomycin inhibits transcription in all forms of life and is highly toxic.

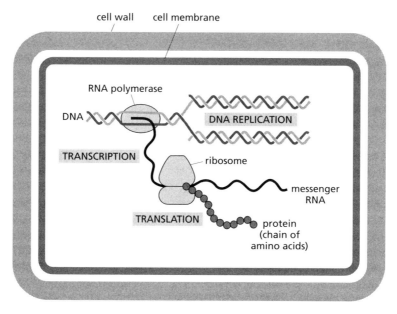

Figure 2.2. DNA replication and gene expression, from the DNA of the chromosomes, via transcription of the messenger RNA (mRNA) by RNA polymerase, to translation of the mRNA into protein on the ribosomes.

As the hunt for new antibiotics heated up, some actinomycete products were found to have useful applications other than to kill bacteria. Rachel Brown and Elizabeth Hazen at the New York State Department of Health in Albany found the first antifungal antibiotic in 1950 and called it nystatin (after New York State). Hubert Lechevalier, a French Canadian who joined Waksman's laboratory as a PhD student, discovered another in 1953, naming it candicidin ("killer of Candida"). *Candida albicans* is a yeast-like fungus that causes thrush, an infection of the mouth or vagina in which the fungus grows as creamy patches, leading to inflammation and tissue damage. These antibiotics kill fungi by interacting with the steroids in cell membranes, making holes through which essential sodium and potassium ions leak out of the cells. Bacterial membranes lack steroids, so they are not susceptible. The antibiotics are used as topical treatments, for example in a vaginal ointment, but are too toxic to be injected or swallowed, because they damage host cells too.

Attempts to treat viral infections and cancers with antibiotics meet with similar difficulties; there is no such thing as a nontoxic antiviral or anticancer drug. The objective is to find compounds with the biggest therapeutic window, so that a dose high enough to inhibit the cancer or virus does not damage the host too much. Viruses consist of DNA or RNA molecules packaged inside protective protein coats. When the nucleic acid infects a host cell, it uses the cell's machinery to replicate, transcribe, and translate the viral genes, so inhibiting a viral infection usually damages the host. But some special attributes of viruses set them apart from mammalian cells, and these can be targeted. The best known example is an enzyme that the human immunodeficiency virus (HIV) uses to copy the RNA molecule in the infectious particle into DNA, which inserts itself into a host chromosome and takes over the cell and its progeny as AIDS gradually develops. Mammalian cells lack this reverse transcription process, and synthetic drugs that block it, such as AZT, are reasonably nontoxic. Attempts to kill cancer cells often exploit the fact that they typically grow and divide faster than most normal cells. Adriamycin, an important anticancer drug, inhibits DNA replication. Actinomycin, which initially failed as an antibacterial antibiotic because of unacceptable toxicity caused by its inhibition of transcription, eventually found a specific use to treat a juvenile kidney cancer called Wilms' tumor. The problem with anticancer therapy is that fast-growing normal cells, such as those in the bone marrow, are likely to be killed too, so a balance has to be struck between trying to stop the cancer and avoiding serious side effects.

Other novel antibiotics found uses in veterinary medicine. The most important is avermectin, which has greatly reduced the ravages of invertebrate parasites such as nematode worms and warble flies in cattle and other livestock. It interferes with chemical signaling by nerve cells of a particular kind. These neurons occur in the peripheral nervous systems of invertebrates, and avermectin causes fatal paralysis by blocking their junctions with the muscles. In mammals, neurons of this type occur only in the central nervous system, where avermectin has difficulty penetrating, so they are unaffected at doses that kill the parasite. Avermectin has also had a large impact on a dreadful tropical human disease called river blindness, caused by a microscopic worm. Spread by bites from a black fly, the parasite infects 18 million people in sub-Saharan Africa and causes blindness if untreated. An annual prophylactic dose of a chemical derivative of avermectin provides protection. Merck has had a policy of making it available free, and 65 million doses were recorded between 1987 and 1996, a proportionately small but welcome return of some of their antibiotic-derived profits to a community that needs help. Later a second benefit emerged: the drug, in combination with another donated by GlaxoWellcome, also cures disfiguring elephantiasis cased by a related parasitic worm.

Microbes produce a range of other compounds used as human medicines for applications that have nothing to do with infectious disease. The statins, made by various fungi, reduce the risk of heart attacks and strokes in susceptible individuals by inhibiting a liver enzyme that catalyzes a step in the production of cholesterol from animal fats in the diet, thereby blocking the route to most of the cholesterol that ends up in the blood (no more than 15% comes directly from cholesterol in food). It is easy to imagine that this ability might have evolved as a response to competition in the soil, because eukaryotic microbes also need to make sterols. More surprisingly, microbes make compounds that suppress the human immune system, a prerequisite for successful organ transplantation. A fungus makes cyclosporin, the first successful immunosuppressant, and others are made by actinomycetes. Why would the soil-inhabiting *Streptomyces hygroscopicus* produce FK506 (marketed as tacrolimus), which docks precisely with a pair of mammalian protein receptors, thus preventing the chemical signaling that triggers the immune response? Perhaps the compound can play some other, as yet unknown, role in the life of the microbe and the effect on the mammal is an incidental consequence. This might seem a rather lame assumption, but it turns out that FK506 has weak antifungal activity and that this depends on interaction with an evolutionary forerunner of the FK506-binding protein of mammals.

Even the most promising drugs may fail as medicines if they do not find their way to the appropriate organ and remain there long enough to do their job. Some antibiotics, although stable in pure solution, break down rapidly in the body. The acid in the stomach is a particular challenge. This was a problem with the natural penicillin that Florey and Chain first gave to patients, which had to be administered by injection twice a day. A huge advance was the production of an acid-resistant chemical variant of the antibiotic that could be taken orally. Some compounds are cleared so rapidly by the kidneys that they never reach an effective concentration in the blood and hence in the tissues. Others are taken out of circulation by binding to serum proteins, or do not penetrate well enough to the appropriate organs to be effective. All of these "pharmacokinetic" considerations, added to problems of toxicity, greatly complicate the discovery of useful antibiotics.

There is another area of antibiotic use that is not subject to such problems. It too has been enormously important for advances in medicine, but indirectly, by increasing our knowledge about how cells work. Natural antibiotics—as distinct from synthetic chemicals—have become adapted over evolutionary time-scales to interact with particular cellular targets, so they can be used to block particular processes in biochemical research. Figure 2.3 shows how vancomycin binds very precisely to the chemical bridges that join the strands making up the growing bacterial cell wall, inhibiting its expansion. If vancomycin is added to a bacterial culture, precursors of cell wall assembly up to the point of inhibition may be isolated, and their structure tells us about the normal pathway. If a particular reaction in a piece of experimental tissue or a microbial culture continues after actinomycin is added, blocking transcription, we can conclude that a stable messenger RNA has already been made and can

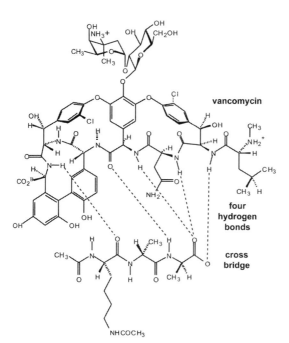

Figure 2.3. Vancomycin docking with its target, the cross-bridge that links adjacent strands of the bacterial cell wall. Hydrogen bonds (dashed lines) connect four atoms of the cross-bridge to specific atoms on vancomycin (one of the oxygen atoms can make two alternative bonds to the antibiotic). (Courtesy of Dudley Williams, University of Cambridge.)

function during the course of the experiment. If chloramphenicol stops a cellular process more or less instantly, active protein synthesis must be occurring to drive the process; if not, we deduce that an enzyme can work without having to be constantly replenished, or that the reaction goes spontaneously with no need of an enzyme to catalyze it. It is hard to imagine the progress of modern biochemistry and cell biology without this wonderful toolbox, and it contains many antibiotics that would be too toxic to use as medicines, greatly adding to the legacy of the Golden Age of antibiotic discovery.

In the next section I describe how pharmaceutical companies went about finding antibiotics and bringing them to market during the Golden Age. It was a difficult and uncertain process. An entry in *The Abbott Almanac* for 1949, early in the game, is telling: "Abbott has isolated more than 3,000 antibiotic-producing organisms and is finding 15 to 20 new ones each month. So far, not one has shown the promise of the mold spore that Alexander Fleming discovered in his laboratory." But there were enough medical and commercial successes for the industry to thrive.

An Antibiotics Industry Develops

In the immediate postwar years, the main U.S. pharmaceutical companies formed two "clusters." In the Midwest were Eli Lilly in Indianapolis, Upjohn in Kalamazoo, Abbott Laboratories in North Chicago, and Parke Davis in Detroit. The other six companies were on the East Coast, three in New York State—Pfizer in Brooklyn (later Groton, Connecticut), Lederle Laboratories in Pearl River, and Bristol-Myers in Syracuse—and three in New Jersey—Hoffmann-La Roche in Nutley, Squibb in New Brunswick, and Merck in Rahway. By the mid 1940s, all of them were heavily involved in producing penicillin, and some were also making streptomycin. Having seen the big profits to be made from successful antibiotics, they were eager to discover compounds of their own.

In 1943, in a unique departure from their usual rivalry, the four Midwest firms pooled their efforts to synthesize penicillin chemically, as a commercial alternative to isolating it from cultures of the fungus. This objective did not get off the ground; penicillin remains to this day a fermentation product, like all the other natural antibiotics except chloramphenicol, which is simple enough to be made economically by chemical synthesis. The four companies continued to share information as they moved into antibiotic discovery. When Parke Davis, in collaboration with a botanist at Yale University, Paul Burkholder, discovered chloramphenicol in 1947, they left the Midwest Group, but the other three companies continued to collaborate right up to 1952, when erythromycin was discovered. Upjohn concluded that this was no more than "a weak penicillin" and did not take it forward. The company went on to discover novobiocin, a much less widely used antibiotic, in 1956, but the other two firms realized erythromycin's potential. In Abbott's hands, especially, it became one of the all-time successes of the antibiotic era and is still enormously important today. Meanwhile, the East Coast group were also having successes: Lederle introduced chlortetracycline, which they called aureomycin, in 1948, and Pfizer launched oxytetracycline as terramycin in 1950. Interestingly, the molecular target of chloramphenicol, erythromycin, and the two variants of tetracycline, like that of streptomycin, is the bacterial ribosome, but the antibiotics represent four different chemical classes and inhibit ribosomes in different ways. They, along with neomycin, another ribosome inhibitor discovered by Lechevalier, laid the main foundations for the success of the Golden Age in the United States.

In Japan, things developed rather differently, perhaps because the pharmaceutical industry was much more fragmented. Japan became self-sufficient for penicillin production by 1948, through the efforts of Hamao Umezawa, a legendary figure in Japanese natural product chemistry, who began working on penicillin in 1944 at the Military Medical School in Tokyo; by 1954 no fewer than 83 companies were involved in making and marketing it.[1] The discovery of new antibiotics in Japan was at first left largely to academic groups. Umezawa, by then at the Institute of Microbial Chemistry in Tokyo, discovered kanamycin in 1957, an important find because it was active against pathogens variously resistant to penicillin, streptomycin, or chloramphenicol. Kasugamycin and bleomycin came later, in the 1960s. Satoshi Ōmura at the Kitasato Institute in Tokyo was the other major academic player. Only later did Japanese industry have important successes, notably in the discovery of natural products with applications outside antibacterial chemotherapy.

How did the new antibiotics industry go about finding and developing these money-spinning drugs? As the companies settled into their task, different departments or divisions were set up for the various stages through which an antibiotic would pass from the soil to the clinic. First came the discovery laboratories, in which soil organisms were isolated and screened for activity. The key to success was novelty, so at an early stage it was crucial to try to eliminate already known compounds, such as tetracyclines, which cropped up all the time. This involved biological tests against a range of microbes, including those that were resistant to earlier antibiotics, as well as chemistry, which became faster and more certain as techniques for determining chemical structure developed. The precise structure of a new compound was crucial, both scientifically and in order to write definitive patents. Biochemical studies of a compound's mode of action were also informative: did it block cell wall synthesis, DNA replication, transcription, translation, or none of the above?

If the new compound was an antibacterial agent, it was important to obtain early evidence that it was effective without killing experimental animals. Usually, a protection assay was done in which a group of mice were given a lethal dose of a pathogen and half of the group was treated with the new drug candidate. If they survived without obvious ill effects, an early hurdle had been cleared. Next came detailed tests of potential toxicity, in animals such as rats, guinea pigs, and dogs, as well as pharmacokinetic evaluation, which might eventually extend to monkeys. Improvements in producing and purifying the antibiotic would also be made, so that enough was available for the procedures that lay ahead, at which point the project would normally pass to a separate department for development. The hope was to identify a compound with enough promise to justify clinical trials; if those were successful, the way would be cleared to scaling up production and eventually to marketing a new product. Let us follow the progress of an antibiotic through some of these stages.

Finding New Antibiotics

The first prerequisite was a supply of soil samples from which to isolate the microbes to be screened for antibiotic production. Sometimes, companies collected them from particular habitats, or from different parts of the world. Marianna Jackson (Figure 2.4), who worked at Abbott Laboratories between 1952 and 2000, told me that they used soil survey maps to help to sample the environment as thoroughly as possible. At Abbott and other companies, employees traveling on business, or even on vacation, would take plastic bags and labels and make a point of collecting a spoonful or two of soil from any likely spots. When they went to a scientific meeting, members of the Abbott discovery group would add a couple of days' collecting onto the trip, going to nearby localities with special features, such as deserts or alkaline lakes in the American West. Occasionally, graduate students were paid to go on collecting expeditions. These in-house efforts were supplemented by purchase of soils from such outside sources as the U.S. Geological Service.

Some interesting hunches were backed from time to time. For example, at one period Abbott reckoned that certain actinomycetes might be more abundant in deeper soil. Local cemeteries would tip them off whenever a new grave was being dug, so that one of the discovery team could drive round with a ladder and clamber down to collect samples. On occasion, a little more faith came into antibiotic discovery. For example, Hamao Umezawa isolated the culture that makes kasugamycin from the Kasuga shrine, founded in 768 AD by the Fujiwara clan at Nara, Japan, and steeped in centuries of tradition. And Satoshi Ōmura, at the Kitasato Institute in Tokyo, collected the soil sample that yielded avermectin, which became an amazing commercial success, from a favorite golf course. He has probably discovered more "splendid

Figure 2.4. Marianna Jackson in the new screening laboratories at Abbott Laboratories, 1983. (Courtesy of Marianna Jackson.)

gifts from microorganisms" (his words) than any other scientist—some 330 interesting compounds of microbial origin. Perhaps Ōmura's skill in finding antibiotics owes something to his early training as a geologist. I once accompanied him on a trip to the Southern Alps, the range of mountains inland from Mount Fuji, and he told me that he loved the outdoors, not only for the beautiful scenery but because he can "tell the structure of land where microorganisms are living" (Figure 2.5). It was a privilege to see this small, dapper figure striding through the swirling mist in a quest for the actinomycetes' secret homeland.

It is difficult to know precisely how much these stratagems really increased the chances of finding a valuable drug candidate, such are the numbers of microorganisms in the environment—one gram of soil can contain 10 billion microorganisms, representing thousands of different species—but they must have helped to spread the net wider. Some later studies confirm this. For example, Boyd Woodruff, the discoverer of actinomycin and streptothricin in Waksman's laboratory, returned to this question after a long and successful career at Merck. In this retirement project, he found differences in the populations of actinomycetes at two depths in the same soil, even at the surface compared with a few centimeters down; in soils of different geological type only a few miles apart; and in the thin layer of soil around plant roots compared with plant-free soil nearby. From fewer than 5500 actinomycetes from diverse habitats he found—in collaboration with the Kitasato Institute—10 interesting new chemical compounds with effects on targets as different as testosterone-requiring prostate cancer cells and a protozoan parasite of chickens.[2] These statistics are much better than those typically recorded from screening huge numbers of actinomycetes isolated more randomly.

Back in the laboratory, the soil would be treated in various ways. It might simply be shaken with water and coarse particles filtered off before the resulting extract was spread on a nutritive agar medium at several different dilutions to give nicely sepa-

Figure 2.5. Satoshi Ōmura (center) with the author and Haruo Tanaka (left) in the Southern Alps, Japan, November 12, 1983. (Courtesy of Satoshi Ōmura, Kitasato Institute.)

rated microbial colonies (Color Plate 1). Representative colonies were picked, usually by hand, though later some companies used automated workstations. In the most sophisticated systems, a television image of the plate was stored in a robot's memory, so that an arm carrying a sterile probe could home in on the positions of colonies that had an appropriate kind of outline or surface contour, typical of actinomycete colonies. In this way, samples of spores were picked up from the target colonies and transferred to culture plates, such as those with 24 separate compartments seen in Color Plate 2A. After these new cultures had grown, the plates were carefully archived in the refrigerator for future use if a promising result was obtained in the screens.

Often, precautions were needed to achieve plates of colonies suitable for picking. Soils contain large numbers of fungi and bacteria. Some motile bacteria can swarm over an entire agar plate in a few hours, and some molds can grow so fast that they cover a plate overnight, swamping the small, compact colonies of the actinomycetes. Therefore, selective procedures were needed to recover the actinomycetes at the expense of the other microbes. The simple sugar glucose, which almost all microorganisms can use as a carbon source, might be replaced by a complex carbohydrate such as chitin or a material such as oatmeal, which most actinomycetes can use and many other microbes cannot, or propionic acid, which actually inhibits most other microorganisms. Sometimes antifungal antibiotics were added to suppress the growth of fungi. Often the soils were air-dried before plating to exploit the desiccation resistance of actinomycete spores. At Abbott, the soil samples were even refrigerated for a couple of years before drying, to kill off the more fragile cells of most bacteria and fungi and favor the survival of actinomycete spores. In other laboratories, bacterial viruses were added to the plates to reduce competition from unicellular bacteria.

A neat trick was to spread the soil extract over a very fine-mesh membrane lying on the surface of the nutrient medium. Unicellular bacteria could not penetrate the membrane. Filamentous microbes could push through the pores, but typical fungal hyphae could not pass, so only the very fine hyphae of the actinomycetes could get through and grow down into the underlying agar. After a first period of incubation, the membrane would be peeled off, revealing colonies of actinomycetes growing on the plate.

Among the actinomycetes, members of the genus *Streptomyces* continued to be the most productive, but some companies focused on other genera of so-called "rare actinomycetes" to try to keep ahead of the game. This approach received more and more emphasis as taxonomists and microbial ecologists discovered new kinds of actinomycetes: there are now approximately 120 actinomycete genera, instead of the five that were enough for Waksman and Henrici in 1943.

Schering Corporation in Bloomfield, New Jersey, chose *Micromonospora* species through a series of fortunate events that George Luedemann has described.[3] In 1957, he joined what was then a relatively small pharmaceutical company. Schering did not have an antibiotic—they were focusing on steroids for the sex hormone and cortisone market and eventually hit the big time with a process for converting cortisone and hydrocortisone into the anti-inflammatory drugs prednisone and prednisolone. They were 15 years behind the big companies in searching for antibiotics. Luedemann had been a graduate student at Syracuse University, New York, where Americo Woyciesjes had been developing special methods for isolating and culturing *Micromonospora* species. Most of his collection had been consigned to the autoclave and destroyed after a change in the research interests of the department, but Luedemann arranged for Woyciesjes to receive a small grant from Schering to set up a tiny laboratory in the basement of his house and send batches of *Micromonospora* cultures to the company for antibiotic screening. Woyciecjes went back to habitats that had previously yielded *Micromonospora* cultures—for example, the saline pools that surrounded Syracuse and had put it on the map as a producer of salt. Woyciesjes supplied Schering with more than 300 *Micromonospora* strains between 1958 and 1960, of which 15 produced apparently new antibiotics. One of them, gentamicin, went on to become a major product for the company. (Note the suffix -micin instead of -mycin, to distinguish this from a *Streptomyces* antibiotic.) Gentamicin was making more than 200 million dollars a year for Schering by 1969. Unfortunately, as in the case of streptomycin, there was litigation: Woyciesjes was not included on the patent or initially rewarded for his part in the discovery of gentamicin until he took the company to court. Why did this happen? When company employees make discoveries their names must appear as inventors on any resulting patents, which are otherwise invalid, but they will have signed away any financial rights when they joined the company, which bears all the legal costs of obtaining and defending the patents, and profits from them. Perhaps there was a (Freudian) slip in Woyciesjes's case, so that he was thought of as an employee when he was not.

An especially cunning method was used to find *Actinoplanes* species, which produce motile spores. John Couch of the University of North Carolina at Chapel Hill discovered this new actinomycete genus in 1950 when he was using the method to isolate aquatic fungi. Animal hairs, pollen grains, or pieces of grass were placed on the surface of water poured over soil samples. The organisms would swim to these "baits" and develop into tiny colonies that could be picked off under a microscope. Teicoplanin, discovered by the Lepetit Company in Milan in 1975, is the best-known antibiotic to come from this genus of actinomycetes.

Table 2.1 shows the total numbers of antibiotics and other biologically active compounds discovered from various groups of microbes up to 2002, and Table 2.2 lists the most useful ones. It is hard to imagine the total number of organisms screened in order to find them. By the end of their isolation program, the Abbott team alone was looking at as many as 100,000 organisms a year, so screening them was a huge operation.

Table 2.1. Numbers of natural products from three groups of microorganisms, up to 2002

Producers	Antibiotics	Other activities[1]	Totals
Actinomycetes	8,700	1,400	10,100
Other bacteria	2,900	900	3,800
Fungi	4,900	3,700	8,600
Total	16,500	6,000	22,500

[1]Including anticancer, antiparasitic, immunosuppressive, and herbicidal agents.

From Bérdy, J. (2005). Bioactive microbial metabolites, a personal view. *Journal of Antibiotics* 58, 1–26.

The tests used to find antibiotics increased in sophistication as time went on. Variations of the agar streak method used by Waksman's laboratory (Figure 1.10), often automated, continued to be used by companies for many years. Usually, some of the liquid from a small-scale culture of the candidate actinomycetes would be spotted onto a culture plate spread with the test organism, or into wells cut in the agar, so many tests could be done on a single plate, and they could be automated. A robot could add the samples, and an automated plate reader could measure the size of the inhibition zones. Later, there was a gradual switch to more sensitive and specific tests, such as interference in a step in cell wall biosynthesis. In searches for pharmacological applications, enzyme inhibition tests could be used, for example in the pathway that leads to cholesterol. Even later, tests were based on miniaturized receptor-binding assays.

A microbe living in the soil produces an antibiotic only under specific circumstances, so choice of the right growth medium for the screens was crucial. Antibiotic production is often a response to a shortage of nutrients—for example, a ready source of energy such as a simple sugar, or an easily available nitrogen source such as ammonia, or a supply of phosphate. Therefore, an abundant supply of any of these nutrients might suppress antibiotic production. Satoshi Ōmura found that adding a special kind of Japanese clay called Kanumatsuchi to the culture medium led to the production of antibiotics that would otherwise have been missed. The clay is used in bonsai culture because it absorbs phosphate and therefore helps to restrict the growth of the tree by limiting its supply of this essential nutrient. It does the same for *Streptomyces,* and the resulting "starvation" signal triggers the antibiotic response.

When new antibiotics were found in the screens, a series of tests had to be performed before a decision could be made as to whether to begin the path to developing the new discovery as a commercial product. Was the antibiotic active at a low enough concentration, and was the range of microbes that it killed what the company was looking for? Was a pharmacologically active compound specific enough to bind to its target but not to interfere with other receptors? Toxicity studies started with *in vitro* assays on cultured cells to see if they were killed. If these gave favorable results, toxicity and pharmacokinetic tests on animals followed. At any stage, hopes might be dashed if the antibiotic spectrum turned out to be too narrow or otherwise uninteresting, if toxicity tests in animals revealed unacceptable side effects, or if the compound seemed to lack activity in the body because it was broken down or excreted too fast.

As the path to development of a new antibiotic progressed, larger and larger amounts of the compound were needed, especially when the stage of animal testing was reached. This, then, was the next challenge.

Increasing Antibiotic Yield

Actinomycetes in the soil grow and undergo their characteristic life cycle on solid substrates, such as particles of decomposing plant and animal debris (Figure 2.6), and antibiotic production typically occurs at the transition from vegetative growth to sporulation, as an adaptive response to competition. The same is true of the penicillin-producing fungus. In order to mimic the natural conditions for antibiotic biosynthesis on

Table 2.2. Natural products of microorganisms and their applications

Natural product[1]	Producer[2]	Biochemical target	Application	Discovering country	Discovering institution
Actinomycin	*S. antibioticus*	Transcription	Anticancer	USA	Academic
Adriamycin (doxorubicin)	*S. peucetius*	DNA replication	Anticancer	Italy	Commercial
Amphotericin (natamycin)	*S. nataensis*	Membrane sterols	Antifungal	USA	Academic
Avermectin	*S. avermitilis*	Invertebrate neurotransmission	Antiparasitic (worms and insects)	Japan/USA	Academic/Commercial
Avoparcin	*S. candidus*	Bacterial cell wall	Growth promotion of farm animals	USA	Commercial
Bacitracin	*Bacillus* species	Bacterial cell wall	Antibacterial	USA	Academic
Bialaphos	*S. hygroscopicus*	Nitrogen metabolism	Herbicide	Japan	Commercial
Bleomycin	*S. verticillus*	DNA replication	Anticancer	Japan	Academic
Candicidin	*S. griseus*	Membrane sterols	Antifungal	USA	Academic
Cephalosporin C	*Cephalosporium acremonium*	Bacterial cell wall	Antibacterial (eye infections)	UK	Academic
Chloramphenicol	*S. venezuelae*	Bacterial ribosomes	Antibacterial	USA	Academic/Commercial
Chlortetracycline	*S. aureofaciens*	Bacterial ribosomes	Antibacterial	USA	Commercial
Clavulanic acid	*S. clavuligerus*	Inhibits penicillinase	Antibacterial (combined with a penicillin derivative)	UK	Commercial
Cyclosporin	*Cordiceps subsessilis*	Lymphocytes	Immunosuppression	Switzerland	Commercial
Epothilone	*Sorangium cellulosum*	Microtubules (cell skeleton)	Anticancer	Germany	Academic
Erythromycin	*Saccharopolyspora erythraea*	Bacterial ribosomes	Antibacterial	USA	Commercial
Fosfomycin	*S. wedmorensis*	Bacterial cell wall	Antibacterial	Spain	Commercial
Gentamicin	*Micromonospora* species	Bacterial ribosomes	Antibacterial	USA	Commercial
Gramicidin	*Bacillus brevis*	Bacterial cell wall	Antibacterial	USA	Academic
Griseofulvin	*Penicillium griseofulvum*	Microtubules (cell skeleton)	Antifungal (athlete's foot)	UK	Commercial
Kanamycin	*S. kanamyceticus*	Bacterial ribosomes	Antibacterial	Japan	Academic
Kasugamycin	*S. kasugaensis*	Protein synthesis	Antifungal (agriculture: rice blast)	Japan	Academic
Lincomycin	*S. lincolnensis*	Bacterial ribosomes	Antibacterial	USA	Commercial

Name	Organism	Target	Use	Country	Status
Lovastatin	*Aspergillus terreus*	Cholesterol synthesis	Cholesterol control	USA	Commercial
Mitomycin	*S. caespitosus*	DNA replication	Anticancer	Japan	Commercial
Monensin	*S. cinnamonensis*	Cell membrane	Growth promotion of farm animals	USA	Commercial
Neomycin	*S. fradiae*	Bacterial ribosomes	Antibacterial	USA	Academic
Nikkomycin	*S. tendae*	Chitin synthesis	Antifungal, insecticide	Germany	Academic
Novobiocin	*S. niveus*	Bacterial DNA replication (gyrase)	Antibacterial	USA	Commercial
Nystatin	*S. noursei*	Membrane sterols	Antifungal	USA	Academic
Oleandomycin	*S. antibioticus*	Bacterial ribosomes	Antibacterial	USA	Commercial
Oxytetracycline	*S. rimosus*	Bacterial ribosomes	Antibacterial	USA	Commercial
Penicillin	*Penicillium chrysogenum*	Bacterial cell wall	Antibacterial	UK	Academic
Polyoxin	*S. cacoi*	Chitin synthesis	Antifungal (agriculture)	Japan	Commercial
Pristinamycin	*S. pristinaespiralis*	Bacterial ribosomes	Antibacterial	France	Commercial
Rapamycin	*S. hygroscopicus*	Lymphocytes	Immunosuppression	Canada	Commercial
Rifamycin	*Amycolatopsis mediterranei*	Bacterial RNA polymerase	Antibacterial (mycobacteria)	Italy	Commercial
Spinosyn	*Saccharopolyspora spinosa*	Neurotransmission	Insecticide (cotton pests, etc)	USA	Commercial
Streptothricin	*S. lavendulae*	Bacterial ribosomes	Growth promotion of farm animals	USA	Academic
Streptomycin	*S. griseus*	Bacterial ribosomes	Antibacterial	USA	Academic
Tacrolimus (FK506)	*S. hygroscopicus*	Lymphocytes	Immunosuppression	Japan	Commercial
Teicoplanin	*Actinoplanes teicomyceticus*	Bacterial cell wall	Antibacterial	Italy	Commercial
Tetracycline	*S. aureofaciens*	Bacterial ribosomes	Antibacterial	USA	Commercial
Thienamycin	*S. cattleya*	Bacterial cell wall	Antibacterial	USA	Commercial
Tylosin	*S. fradiae*	Bacterial ribosomes	Growth promotion of farm animals	USA	Commercial
Vancomycin	*Amycolatopsis orientalis*	Bacterial cell wall	Antibacterial	USA	Commercial
Virginiamycin	*S. virginiae*	Bacterial ribosomes	Growth promotion of farm animals	USA	Commercial

[1]**Bold type** indicates actinomycete products; normal type indicates fungal products; *italics* indicate products from non-actinomycete bacteria.

[2]*S.* stands for *Streptomyces*.

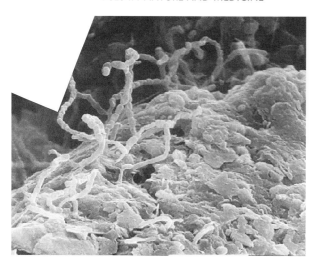

Figure 2.6. Scanning electron micrograph of *Streptomyces* growing on a particle of organic material in the soil. Note the spore chains developing in an air space. (Courtesy of Elizabeth Wellington, University of Warwick.)

solid surfaces, the first large-scale production of penicillin was in vessels containing a shallow unstirred liquid medium, and the mold formed a skin-like growth floating on the surface. Alternatively, large trays of solid medium were used. Only later was it possible to produce antibiotics by growing the organisms—whether actinomycetes or fungi—much more efficiently in giant stirred tanks of liquid medium. This technology was familiar to fermentation companies used to making beer or industrial alcohol with yeasts, which are naturally adapted to living in liquid environments, but antibiotic production in liquid had to be forced on the filamentous actinomycetes and molds by changing the medium or the physical conditions in the fermenter.

The amount of antibiotic made by an actinomycete is like other quantitative characters in domesticated plants or animals, such as the yield of grain by a wheat plant or the thickness of the woolly coat of a sheep. It is controlled both by environmental factors and by the genetic make-up of the organism. (It is just the same for human attributes such as stature, or physical or mental skills: the beauty of genetics is that principles worked out with one group of organisms usually apply to all.) To optimize the environment, a group of microbial physiologists and fermentation experts experimentally varied the culture medium, the temperature, the pH, and the aeration of the culture.

In parallel with these fermentation studies, geneticists improved the strain genetically. Right from the mid-1940s, mutagens such as X-rays, ultraviolet light, and a range of chemical compounds were used to increase the chance of finding mutations that increased the organisms' productivity. The survivors of a mutagenic treatment were screened laboriously to find those that were improved. Robots and plate readers could do parts of this work, as in the hunt for antibiotic production by a newly isolated culture. Improvement of the strain was a slow process, because many genes with cumulative effects control the amount of antibiotic made, and changing just one or two genes rarely increases the productivity drastically.

Sometimes the need to optimize both genetic and environmental factors resulted in a degree of friction between the geneticists and the fermentation team. After the latter had carefully optimized the fermentation for a given strain, along came the next mutant, which made more antibiotic under small-scale laboratory conditions but behaved differently from its predecessor during scale-up for growth in large tanks. Optimization of the fermentation would then have to be repeated.

The amount of antibiotic made in some of the best fermentation processes, such as those for penicillin and the tetracyclines, was increased over a 20-year period as much as 1000-fold. Huge fermentation tanks contained as much as 50 g of antibiotic per liter of medium. Very little of this improvement over the original productivity was the result of deliberate changes, either in the genetic make-up or in the physiology of the organisms, because the necessary knowledge base that would develop during 40 years of subsequent research was not available during the Golden Age. And in any case, the strategy of random mutation and screening, however laborious and empirical, was working well enough—and all the companies were using it, with little likelihood of one company's suddenly stealing a march on the others—so there was no imperative to find a more rational strategy.

After the discovery department of the company had done its job, a promising antibiotic would be passed on to the development section. Already, tests for efficacy, toxicity, and pharmacokinetics would have been done, the strain and fermentation process would have been put through the first stages of improvement, and chemical isolation would have been addressed. Another, less serious, issue would also have been solved: a name for the new compound, a separate matter from the trade name that would be used eventually if the material came to market.

The names applied to the first antibiotics discovered by Waksman's group—actinomycin, streptothricin, and streptomycin—reflected generic names that had been applied to the microbes that made them, just as penicillin derived its name from the *Penicillium* fungus from which it came. Soon the genera of actinomycetes had been exhausted, and antibiotic names were derived from the specific names of the producers, for example aureomycin from *Streptomyces aureofaciens* and erythromycin from *Streptomyces erythraeus,* denoting the gold or red color of the cultures. Other names reflected some aspect of the chemistry of the antibiotic, such as chloramphenicol, which contained chlorine atoms and a phenolic carbon ring and is an alcohol, or of its target, such as candicidin, which killed *Candida*. Most names were mundane, but others had a jokey flavor. When Alexander Argoudelis, a scientist of Greek descent at the Upjohn Company, needed to name a new antibiotic, he called it zorbamycin. He named another compound melinacidin after Melina Mercouri (1923–1994), who made her name in the 1960 movie *Never on Sunday* and went on to be Greece's Minister of Culture. Perhaps the worst pun was lipiarmicin, a name given by Lepetit to an antibiotic made by one of their *Actinoplanes* cultures. How was it chosen? The organism was isolated on February 29 (try saying "leap-year-micin" with an Italian accent).

Into Production

The key decision whether to put a new antibiotic into production depended on reliable information on its likely clinical safety and effectiveness. Although animal studies are a good preliminary indicator, tests on human volunteers are eventually needed.

Such clinical trials have become enormously more comprehensive and costly since the early days of antibiotic discovery.

In Phase I, safety and dosage are the chief concerns, safety being assessed by tests on the functions of vital organs such as the kidneys and liver as well as by more general examinations. Then, in Phase II, effectiveness is the main criterion, coupled with careful checks for side effects. Phase III represents the most nail-biting stage of the whole development path. Clinical benefit of the new drug is compared with that of existing treatments, and adverse reactions from long-term use are monitored. Phase III trials are especially expensive, because they involve treating thousands of patients, for perhaps 3 years. If some adverse reaction crops up, it may be judged serious enough to stop the drug in its tracks. The ratio of benefit to side effects for an anticancer drug may allow a potentially life-saving compound to go forward even with inevitable toxicity; this would not be tolerated in an antibiotic that showed improved activity against a bacterial infection, compared with its predecessors, but not enough to justify its use in the face of possible side effects, however slight. So, after years of work, a large outlay of research effort and expense may have to be written off.

A good outcome in the clinical trials was a signal to continue to optimize fermentation conditions in the pilot plant and on into production. This involved a lot of further scale-up work, because a medium that is good for making an antibiotic on a small scale may be either unsuitable or far too expensive for use in a fermenter containing thousands of liters of liquid, extending through several floors of a large building (Figure 2.7). Sources of carbon, nitrogen, and energy must be inexpensive for large-scale production. Instead of highly refined sugars or amino acids, the fermentation media contained fish meal, soya flour, oils of various kinds, or corn steep liquor—a nutrient-rich byproduct of corn processing that became famous in the early days of penicillin fermentation because of its huge impact on antibiotic productivity.

Like nursing stations in an intensive care unit, modern fermentation tanks are extremely sophisticated. The condition of the "patient" needs to be monitored 24 hours a day for the week or more that the fermentation lasts. The tank contains probes for dissolved oxygen concentration in the medium, carbon dioxide leaving the culture as a criterion of respiratory activity, the pH, the supply of nutrients—often supplemented continuously by the equivalent of a surgical drip, because, if easily metabolized nutrients are abundant, the organism typically produces little or no antibiotic. And of course the concentration of the antibiotic accumulating in the "beer" has to be monitored by taking samples at regular intervals and assaying them by either chemical or biological tests. The contents of the vessel have to be stirred by huge paddles, air has to be forced into the culture to supply oxygen, and then the heat produced from all this mechanical energy going in and from the metabolic activity of the microbe has to be removed by cooling coils.

Assuming everything goes according to plan—and occasionally it may not, perhaps because infection by a virus wipes out most of culture and reduces the antibiotic yield catastrophically—the liquid can be pumped off and the antibiotic recovered and purified by whatever process its chemical properties dictate. This is simplest if the antibiotic can be extracted into a liquid organic solvent that does not mix with water, so that the desired compound can immediately be separated from most of the other materials in the "beer" by shaking the liquids together and allowing them to

Figure 2.7. Industrial-scale fermenters (132 cubic meters; 35,000 gallons) at Abbott Laboratories, North Chicago. (A) Installing a new fermenter. (B) View inside a fermenter, showing the impeller that stirs the contents. The worker in a hard hat (arrow) gives an indication of scale. (Courtesy of Janet DeWitt, Abbott Laboratories.)

separate into layers. The solvent is then evaporated off to yield the antibiotic in solid form for further purification. If not, more sophisticated methods, such as adsorption onto and subsequent removal from special resins, must be used. After what can be a long, multistep process, the antibiotic is judged to be pure enough—a standard that is rigorously defined in the U.S. Food and Drug Administration license or equivalent certification of the company's practices—to be formulated as tablets, capsules, or an injectable liquid for shipment to the pharmacy or clinic.

Acquired Antibiotic Resistance

Despite the discovery of so many useful antibiotics by the increasingly imaginative screens employed to find them, there has been a continuing need for new ones. The problem of acquired antibiotic resistance has exercised physicians, academic scientists, pharmaceutical companies, and the media ever since pathogens started to become resistant to penicillin and streptomycin, which happened almost as soon as they were introduced into medicine. Bacteria have adapted over eons to overcome the defenses of their hosts. Their chief weapon is the sheer numbers they can muster, so that even very rare mutations have a chance of occurring whenever they are needed. A simple rod-shaped bacterium can divide every 30 minutes or less, so a single cell growing unrestricted for 24 hours could theoretically give rise to more than 100 million million progeny. All the evidence indicates that the antibiotic does not *induce* antibiotic-resistant mutations—it allows the selective survival of mutants that arise by chance—although every so often the question of adaptively induced mutation (Lamarckian evolution) crops up again.

By far the biggest problem, though, is not mutation but the ability of pathogens to acquire resistance genes from another member of the same species, or even from a different species, by mating with it. Where did such resistance genes originate? In 1970, Margaret and James Walker, pioneers of the study of streptomycin biosynthesis at Rice University in Houston, made the far-sighted suggestion that the resistance genes came from the antibiotic-making actinomycetes, which have to protect themselves from their own weapons. They do this in several different ways.

A neat means by which bacteria resist an antibiotic is to modify the target for the drug inside the producing cell. This is how the producer of erythromycin avoids suicide. An enzyme inserts a methyl group at a crucial spot on the ribosomal RNA, thereby preventing erythromycin from binding to the ribosomes and inhibiting protein synthesis. Another way is to pump the antibiotic out of the cell as it is made, thus preventing it from attacking the target (the ribosomes, in the case of tetracycline). If any tetracycline reenters the cell, it is immediately pumped back out, so the ribosomes of the producer do not need to be resistant. A third strategy is to attach a protecting group, such as phosphate or acetate, to an early precursor of the antibiotic. This group stays on the molecules as they move along the biosynthetic pathway until, as the finished antibiotic leaves the cell, the protecting group is removed. Any antibiotic reentering the cell is detoxified by readdition of the phosphate or acetate and reexported. This is how *S. griseus* protects itself against its own streptomycin.

Examples of all these resistance mechanisms are found in the pathogens that have become resistant to the corresponding antibiotics. Raoul Benveniste and Julian Davies, working at the University of Wisconsin, Madison, showed in 1973 that actinomycetes contain many antibiotic-modifying enzymes uncannily like those of the pathogens, so it seems inescapable that the antibiotic producers are the ultimate source of the resistance genes.[4] As they pointed out, there is no reason to think that these genes were transferred directly from the soil-inhabiting actinomycetes to the pathogens. Almost certainly there have been many intermediate carriers in the chain of gene transfer, including harmless relatives of the pathogen.

Enzymes that break down an antibiotic into harmless products can also confer resistance. The most famous example is the penicillinases, which bacterial pathogens use to combat penicillins. Ernst Chain published a letter on penicillinases in the same month, December 1940, in which the Oxford group described their first clinical use of penicillin. The origin of penicillinases is not known; they are not involved in self-resistance by the organisms that produce the antibiotic. The *Penicillium* mold does not need a self-protective mechanism because there is no target for penicillin in its cells, and actinomycetes that make penicillin protect themselves in other ways. In any case, there would be no selective advantage in producing an antibiotic as an aid to competition in the soil, only to break it down again.

Many actinomycetes actually produce compounds that inhibit the activity of the penicillinases made by other organisms, and one, clavulanic acid, has been combined with a penicillin derivative, which would otherwise be broken down by resistant pathogens, to make a highly successful drug, Augmentin. The clavulanic acid inhibits the penicillinase of the pathogen so that the penicillin can kill it. Commercial development of Augmentin as a combination drug by what was then Beechams (now part of GlaxoWellcome) is a real success story. It involved a lot of research, including choosing a special chemical derivative of penicillin that stays in the tissues for the same length of time as the clavulanic acid, so that they are always present together at appropriate concentrations.

Any use of an antibiotic inevitably selects for resistance, but some practices have been implicated in speeding up the process. Particularly controversial has been the widespread addition of antibiotics to animal feed in intensive poultry, pig, and cattle rearing. It was found in the late 1940s that adding a small amount of tetracycline to the feed increased the growth rate by a small but significant amount. Later, other antibiotics were found to have similar effects. Some estimates put the increase at about 10 percent, though it may be as small as 4 or 5 percent. But this was enough to persuade farmers to adopt the practice, and a huge market developed, accounting for up to half of all antibiotics used. The mechanism of growth promotion is not always clear. It might result from the killing of pathogens causing low-level infection in the animals or birds, although the concentrations of antibiotics in the feed are so low that this is questionable.

The way monensin works in cattle is better understood: it selectively inhibits certain kinds of microorganisms in the gut. The rumen is a natural fermenter in which microbes play an essential role in degrading the hard-to-digest plant material that forms much of a cow's diet. They break down the cellulose of plant cell walls, which higher animals cannot digest, into simple sugars. These sugars are normally metabolized by splitting each molecule, which contains six carbon atoms, into two chains of three atoms each and converting these to organic acids such as pyruvic acid. However, certain bacteria in the rumen tend to split the three-carbon chains into acetic acid, with two carbons, and methane, with one. The methane cannot be assimilated, and the cow belches it into the environment, where it is one of the greenhouse gases. So we have a double problem—the cow loses nutritive value from its diet and at the same time pollutes the atmosphere. Monensin tends to inhibit the methane-forming bacteria and leave the coast clear for the useful microbes in the rumen.

Adding monensin to the feed is not really a problem, because neither it nor chemically related antibiotics are ever used in medicine, so there is no selection for dangerous antibiotic resistance genes. Use of some other antibiotics is a different matter, and there have been successive reports, starting with the Swann Report in the United Kingdom in 1969,[5] recommending that medically important antibiotics and their relatives should be outlawed as growth promoters. There were later reports with similar messages, such as those sponsored by the World Health Organization in 1994 and 1997. Some measures were put in place in the United Kingdom and continental Europe. Avoparcin, a member of the same chemical class as vancomycin, was banned in 1997 from animal husbandry in the European Union, where it had been sold as a growth promoter (it was never marketed in the United States), in case its use spread vancomycin resistance in human pathogens. This could be disastrous, because vancomycin is the treatment for infections caused by gram-positive bacteria that have become resistant to all other antibiotics. There has been no legislation against agricultural applications of antibiotics in the United States. In May 2002, an organization called the Alliance for the Prudent Use of Antibiotics (APUA), which was set up in 1981 to publicize the dangers of misusing antibiotics, made a series of recommendations. Since then, legislation has been repeatedly introduced, but the outcome remains to be seen.

The whole issue is complex because of the difficulty of evaluating the real risks. Sebastian Amyes, Professor of Microbial Chemotherapy at the University of Edinburgh, concluded in 2001[6] that, despite the lapse of 30 years since publication of the Swann Report, the extent to which agricultural applications of antibiotics increased the build-up of resistance in clinically important bacteria was by no means resolved, although there were some telling findings. Stuart Levy, Professor of Molecular Biology and Microbiology at Tufts University School of Medicine, Boston, and a prime mover in the establishment of APUA, described in an excellent 2002 book[7] how antibiotics in chicken or pig food could rapidly select for antibiotic resistance in the harmless *Escherichia coli* that inhabits the intestines of people living near the farms. After the withdrawal of avoparcin from agricultural use in the European Union, the incidence of antibiotic-resistant bacteria in animals fell sharply, with a parallel decrease in humans. Recently, a very detailed study was made of the consequences of discontinuing all use of antibiotics for growth promotion in pigs and chickens in Denmark, a measure that was implemented voluntarily by the industry in 1998.[8] There was a dramatic fall in the numbers of resistant bacteria in food animals but little change in the incidence in hospitals. The damage that environmental use of antibiotics can cause may live on long after withdrawal of the antibiotics from farm use.[9] Interestingly, although the Danish study showed a small decrease in average feed efficiency, this was offset financially by saving the cost of the antibiotics.

Another potent factor is the veterinary use of antibiotics to treat actual infections in herds of farm animals, in pets, or in pens of salmon attacked by a pathogen called *Aeromonas salmonicida,* which kills them in large numbers in crowded fish farms. Such therapeutic uses require 10 to 100 times higher doses of antibiotics than those added to the feed to increase growth rate, even if the total used is only about one fifth of all nonhuman usage. There have been some successes in reducing this load; for example, vaccines against *A. salmonicida* have been found to be much more effective than antibiotic treatment, which might therefore be largely phased out.

Amyes pointed out that situations vary enormously in different parts of the world. In India, for example, antibiotics are too expensive to be used by farmers, even therapeutically, never mind as growth promoters, yet antibiotic resistance in human pathogens is extremely high. He blames this on the widespread use by rural people of antibiotics at levels that, while far too low to cure their infections, are all that many of them can afford. In some countries, potentially life-saving antibiotics are sold over the counter to treat trivial, self-diagnosed infections, another extremely undesirable misuse of precious drugs. But the developed world has no right to be complacent. The overprescribing of antibiotics in medicine, in situations in which they are not likely to be effective, is widespread. A 1992 U.S. survey concluded that well over 10 million prescriptions, perhaps one third of those written for antibiotics, were for children diagnosed with virus infections causing common colds or bronchitis, which rarely respond to antibiotic treatment. In fact, Levy estimates, as much as half of all human antibiotic usage in the United States is unnecessary or inappropriate.

The burning need for novel antibiotics is to fight the life-threatening pathogens that have acquired, not one resistance gene, but a suite of genes that protect them against almost all the antibiotics that formerly controlled them. The most notorious example is methicillin-resistant *Staphylococcus aureus* (MRSA), the cause of potentially fatal hospital-acquired septicemia that is hard to eradicate from surgical wards and operating theaters. This "superbug" even became an election issue in the United Kingdom in 2005, when the Conservative party tried to blame its increase on the Labour government for poor cleaning of National Health Service hospitals. The name MRSA comes from a penicillin derivative, methicillin, that was introduced in the 1960s to combat the penicillinases of resistant pathogens, but it would be better to call it multiresistant *S. aureus,* because it is resistant to several antibiotics besides methicillin. Vancomycin has mostly held it at bay, but there is a danger that this antibiotic too will be rendered ineffective. In July 2002, the first truly vancomycin-resistant strain of MRSA was identified in Michigan, a chilling reminder of the need for new antibiotics. Other equally severe threats are posed by vancomycin-resistant *Enterococcus* infections after abdominal surgery, and by multidrug-resistant gram-negative pathogens belonging to the genera *Acinetobacter* and *Pseudomonas*, which can cause life-threatening pneumonia.[9]

There is a clear need to increase awareness not only of the threats posed by acquired antibiotic resistance but also of the sometimes simple procedures that can reduce them. For instance, in British hospitals the role of Matron was abolished in 1969, and many see this as a cause of the declining standards of hygiene that have allowed MRSA to flourish. Matrons are now being brought back, and one of their tasks will be to persuade hospital doctors, who should know better, to swab their hands between patients when doing ward rounds.

Routes to New Antibiotics

Even with the best procedures for hindering the rise of antibiotic resistance, the pathogenic bacteria will fight back when they are exposed to antibiotics, so new ones will eventually be needed. In the last quarter-century, very few new natural antibiotics were introduced into medicine. Figure 2.1 shows several names in that period, but they are almost all immunosuppressants, antiparasitic or anticancer drugs, and even a herbicide. The original strategy of isolating microorganisms from the soil and screening them for antibiotic activity had become subject to a law of diminishing returns. Perhaps all the good antibiotics had been found, and totally new approaches to finding new compounds were needed.

In the late 1930s, before the antibiotics era, the synthetic sulphonamides were valuable antibacterial drugs for the treatment of gram-positive infections, but they were eclipsed by penicillin. Later, the success of the actinomycete antibiotics ensured that natural products were the main focus for new antimicrobial drugs for many decades. Synthetic chemistry continued to play a role, but only rarely did it provide new drugs by *de novo* synthesis. The quinolones, first described in 1962 as a chance byproduct of attempts to synthesize an antimalarial drug, were the big exception. The best known is ciprofloxacin, or Cipro, which became famous in 2001 during the anthrax scare in the United States, because it is especially effective against this pathogen. Typically, chemistry was used to modify natural products, in an approach called semisynthesis. This was absolutely crucial for penicillin, and the related antibiotic cephalosporin, to improve their pharmacological properties and to overcome the ravages of the penicillinases, but much less so for actinomycete antibiotics, which were mostly used in the form in which nature provided them.

One answer to the problem of finding new antibiotics was to return to chemical synthesis. This received a shot in the arm in the 1990s with the invention of "combinatorial chemistry," in which robots make thousands of novel chemical structures. It convinced many in the pharmaceutical companies that the time was ripe to return to synthetic chemistry as the primary source of drugs, and one by one they stopped screening microorganisms. An alternative approach was to try to make "unnatural natural products" by artificially manipulating the ability of the actinomycetes to make antibiotics. This became possible only with a detailed knowledge of the genetics of streptomycetes and the availability of tools for manipulating their genes, as we shall see in Chapter 8. To reach that point, I now follow a story that began in the 1950s, when microbial genetics was a young science, and I began my own research career with a project to try to find out about *Streptomyces* genetics.

3

Microbial Sex

Microbial genetics goes back at least 75 years.[1] In the late 1920s, Bernard Dodge, at the U.S. Department of Agriculture in Washington, DC, and later at the New York Botanic Garden, worked out the sexual cycle of a pink bread mold called *Neurospora crassa* and learned how to make crosses. In the 1930s, Carl Lindegren at the California Institute of Technology made the fungus a good laboratory subject for genetics, isolating mutants and mapping them on the chromosomes. This laid the foundations for George Beadle and Edward Tatum, at Stanford University from 1940, to adopt *Neurospora* for their Nobel Prize–winning work on the direct relationship between genes and enzymes.[2] Later, other fungi, including the mold *Aspergillus nidulans*, and especially brewer's yeast, *Saccharomyces cerevisiae*, became important models for studying genetics.

The genetics of fungi was founded on knowledge of their life cycles, which allowed the sexually produced progeny of crosses to be analyzed in much the same way as in plants and animals. In all these organisms, sex brings together a complete set of genes from each parent and these are then sorted out into new combinations. The genetics of bacteria developed very differently. In them, the processes that give rise to novel combinations of genes, equivalent to sex, were discovered only because of their genetic consequences: no one saw them happening.

When I began my scientific career in 1954, bacterial genetics was still in its infancy but already appeared bizarre compared with the genetics of plants and animals. Exchange of genes between members of a bacterial species resulted from any of three different processes, all distinct from sexual reproduction in fungi. One was a transfer of genes through the culture medium in a process called transformation, first discovered in the pneumococcus as early as 1928 but only understood in 1944 when the mysterious "transforming principle" in the culture fluid of the bacteria was shown to be DNA. Then, in 1946, *Escherichia coli* was found to exchange genes by a process that, as it later transpired, involved a strange form of mating in which pairs of bacterial cells begin to copulate and transfer DNA, only to have the process interrupted part way through the mating. In 1952, a bacterial virus (a bacteriophage) was reported to mediate gene reassortment in *Salmonella typhimurium,* in a process named transduction. In spite of the differences between these bacterial processes, they all resulted in transfer of incomplete sets of genes from a donor cell into a recipient, not the regular fusion of complete gene sets characteristic of fungi and higher life forms.

Bacteria belong to a group of organisms called prokaryotes, separated by more than 2 billion years of evolution from fungi, plants, and animals—the eukaryotes. Eukaryotes are defined as having a nucleus surrounded by a membrane that separates its contents (notably the chromosomes) from the rest of the cell, whereas prokaryotes lack a membrane-bound nucleus. If the actinomycetes really did fall between these two major subdivisions of life, they should be a happy hunting ground for novel genetic phenomena that might be different from anything already described and might throw light on the evolution of sex in higher life forms. Did the actinomycetes even exchange genes at all? This chapter describes how this question was answered as *Streptomyces* genetics began.

Getting Started in Science

My elder sister and I had grown up with a love of plants and animals, fostered by my mother, who taught English and French but was passionately interested in natural history as a hobby. While we were growing up on the edge of the small town of Havant, near Portsmouth on the south coast of England, she took us for walks in the almost unspoilt countryside that started on the edge of town. My interest in plants and animals was reinforced when my sister talked the local farmer into letting us help out on weekends and during the school holidays. We milked the cows, brought in the harvest, and generally enjoyed the rural life. This must have started in 1943, because I remember learning from the head cowman, who had just heard it on the radio, that the June 1944 D-Day landings in Normandy had begun as we brought in the cows for milking one day, and we had been volunteering on the farm for a year or so by then.

Science subjects became my favorite lessons in school, partly because we had excellent teachers. In 1946, we moved to the north of England when my father, who was in the Inland Revenue Service, was transferred from Portsmouth to Manchester (he had fought in the First World War and was too old for the Second, so his contribution to the war effort was to keep the government's tax stream flowing). Lymm Grammar School in Cheshire, just south of the Manchester Ship Canal, was not noted for academic achievement. A small coeducational secondary school, it was usually second or third choice for bright pupils from the village of Lymm, who were creamed off by more high-powered single-sex schools in the nearby towns. However, my sister and I were keen to go to the local school so that we could cycle home for lunch instead of suffering school dinners as we had at our previous school, which had been an hour's bus ride away.

The school owned some housing, which was very scarce just after World War II, so it could attract some first-class teachers when they were demobilized. Our biology teacher, J. K. ("Jake") Newman, was one of them. He had all the attributes of a charismatic teacher, including quirks such as long swept-back hair and an unfamiliar accent (he came from Bristol in the southwest of England), as well as anecdotes about life in the tank corps, combined with an obvious mastery of his subject. As he introduced us to the frontiers of biology, it seemed to me that the life sciences were at a point where anyone with enough motivation could make a contribution, whereas the cutting edges of physics and chemistry were accessible only to a select few in the vanguard of research. But it was Jake's enthusiasm that really drew me to biology as an academic pursuit: as so often happens, a gifted teacher catalyzed a career choice.

At around that time, I saw a film at the local cinema about the Agricultural Advisory Service. It documented a day in the life of a scientist collecting samples of soil and diseased crops from farms and going back to the laboratory to diagnose the problem and advise the farmer how to deal with it. It seemed a great occupation and I realized that I wanted a career combining biology with the open-air life I had enjoyed on the farm. To prepare for this, I was advised to study for a degree in pure science and then take a postgraduate diploma in agriculture. So, when I went to the University of Cambridge in October 1951, I enrolled for botany, zoology, and geology with a minor course in biochemistry, a combination designed to underpin a career in agricultural research. When I reported my choices to my school chemistry teacher, he commented in a letter that biochemistry was a good choice because "it will keep you *au fait* with the antibiotics field." I barely knew what he was talking about and certainly had no idea of the prescience of his remark, even though I had personal experience of penicillin as early as 1947, when I was given it by injection twice a day just in time to cure a grossly swollen septic finger that would have stopped me taking my first important public examinations (the so-called School Certificate). I also heard about streptomycin in a BBC radio news item that same year, in which Waksman's work on *Streptomyces* was announced, but it had not made any special impression on me.

At Cambridge, I soon became hooked on laboratory science and forgot about the idea of an outdoor career. I found genetics the most exciting topic, partly because, although we had covered a wide range of biology in school, it was a revelation to learn about the wonders of heredity. Harold Whitehouse lectured in genetics in the Botany School. He had worked on *Neurospora*, as well as being an expert in field botany, especially mosses (Figure 3.1). He was soft-spoken and apparently tentative at first encounter and was not a charismatic teacher, but his lectures were comprehensive, superbly organized, and always brought in the latest research: once again, a gifted teacher had made a convert. So, after graduating in 1954, I was looking for an interesting topic in genetics for a PhD.

The Botany School played a special role in teaching and research in genetics in Cambridge in the 1950s. The university had created a Professorship of Genetics in 1912, the first in the United Kingdom, and from 1943 Ronald Fisher, the brilliant statistician and population geneticist, held the chair and headed the genetics department. He had established a final-year undergraduate course in genetics but it, and the research topics in the department, were heavily slanted to the quantitative end of

Figure 3.1. Harold Whitehouse collecting mosses at Easter 1950 in the English Lake District. (Photograph by Pat Whitehouse; courtesy of Richard Savage, University of Cambridge; reproduced by permission of Jane Cooper and Anne Whitehouse.)

the subject and accessible only to those with an advanced qualification in mathematics. In contrast, more "biological" genetics had been built up in the Botany School, first by David Catcheside and then, after his departure for Australia in 1951, by Harold Whitehouse. I therefore stayed in the Botany School (Figure 3.2) for my graduate studies.

Harold correctly predicted that microbial genetics was on the verge of a tremendous explosion, based on Watson and Crick's solving of the structure of DNA the year before, which Harold had incorporated into his lectures at the first opportunity. He pointed out the huge advantages that microbes, with their rapid growth rates and large population sizes, were showing as experimental material for studying the way that genes are organized and expressed; these were the very attributes that Beadle and Tatum had recognized. It was Lewis Frost (Figure 3.3), who was on the faculty in Harold's section of the Botany School, who determined my scientific destiny. He was not a typical Cambridge academic. He had been an educator at an army prison for an extended period of National Service, and his passion was the Goon Show, the crazy 1950s BBC radio series. He would recall the latest episode the next day with "Nobby" Clarke, the genetics technician, who then depicted the characters in wonderful chalk cartoons on the laboratory blackboard. But Lewis was also passionate about genetics. He told me that the actinomycetes might be intermediate between

Figure 3.2. The Cambridge Botany School, June 2005. This is the Rockefeller wing, opened in 1934 as an extension to the 1904 building (seen in the background) to accommodate plant physiology, mycology, and paleobotany. From the late 1930s, with the arrival of David Catcheside, the paleobotany laboratories on the first floor were reassigned for genetics. (Courtesy of Nick Hopwood, University of Cambridge.)

bacteria and fungi and therefore would be well worth studying genetically. By chance, then, these microbes were back in my mind after a gap of 7 years since I had first heard that news item about the discovery of streptomycin.

My reading of Waksman's 1950 book,[3] which Harold lent me as a starter, made it very clear that knowledge of the life cycles of the actinomycetes was very sketchy, and that there had been no critical study of their genetics. There was a chapter on "Varia-

Figure 3.3. Lewis Frost (left) with Mike Ames rescuing rare plants before cliff stabilization work at the Avon Gorge, Bristol, U.K., around 1980. (Reproduced by permission of Bristol United Press; supplied by Phillip Garland, University of Bristol.)

tions and Mutations," but it was one of the shortest in the book. One statement was telling: "[the Russian microbiologist, Anatolii Evseevich] Kriss . . . recognized four types of variation—morphological, cultural, physiological and applied." Such distinctions were largely arbitrary and are relevant to use of the variants in industrial applications: they did not reflect fundamental properties of the organisms. I was excited. I had read the book with some trepidation, afraid that many of the secrets of actinomycete genetics might already have been revealed, but I need not have worried: the field was wide open.

Microbial Genetics in 1954

Looking back, after half a century in which genetics has been dominated by studies on microorganisms, it is salutary to see where microbial genetics had arrived by the time I graduated in botany in 1954. Fungal genetics was developing but was still a small endeavor compared with the genetics of plants and animals. Catcheside had published a book in 1951 that covered the whole of microbial genetics in 220 pages, with just one chapter on bacteria.[4] In it, he underlined the primitive state of knowledge about bacterial genetics: "The simplest assumption is that the bacterial cell contains genes. . . . [but they] could all be separate from one another or joined in strings as in chromosomes [of fungi] or joined in some other conformation such as a flat plate."

The Microbial Genetics Bulletin[5] was another token of the embryonic state of microbial genetics. It was founded in 1950 by Evelyn Witkin at the Cold Spring Harbor Laboratory on Long Island, New York, one of the most important institutions in the development of modern genetics. It was a slim newsletter (the first issue ran to 17 pages), distributed every 6 months to an international mailing list that began with just 78 names. It aimed to cover advances in the genetics of both eukaryotic microbes and bacteria and grew to fill a genuine need. By the 10th issue there were 400 names on the mailing list, and a regular feature listed publications in microbial genetics, which, Witkin pointed out, were widely scattered and often in journals unknown to many in the field. The *Bulletin* continued until 1979, by which time there were flourishing journals entirely or largely devoted to genetic research on microbes.

It was Joshua Lederberg, a medical student with Francis Ryan at Columbia University in New York and then seconded to the laboratory of Edward Tatum, by now at the Rockefeller Institute, who really launched bacterial genetics. In 1946, Lederberg and Tatum described a revolutionary new way to detect the equivalent of sex in bacteria, using a laboratory culture of *E. coli*. Instead of looking at bacteria down the microscope, as bacteriologists had done for decades, and at best seeing cells coming together in suggestive ways but never being able to prove that this represented anything other than a chance contact, Lederberg detected the genetic consequences of sexual reproduction by its unique ability to generate novel combinations of the genes of the parents. He could show unambiguously that *E. coli* exchanged genes, even though the process involved only a tiny fraction of the population and could not be seen. He used a class of mutants called auxotrophs, which Beadle and Tatum had exploited in their biochemical work with *Neurospora*.

Wild-type cultures of *Neurospora* and *E. coli*, like most microbes that live in habitats poor in ready-made nutrients, can synthesize all of the hundreds of molecules they need from just a few starting materials. These include a source of carbon such as the sugar glucose, nitrogen in a simple form such as nitrate or ammonia, and a series of inorganic salts. They do not need "dietary supplements" such as vitamins or amino acids, as humans and other animals do. The organisms make all these compounds by biosynthetic pathways in which each step, from the earliest precursor to the finished vitamin or amino acid, is catalyzed by an enzyme specified by a gene: proof of this "one gene, one enzyme" rule was Beadle and Tatum's major achievement. Therefore, if the gene is damaged—mutated—its enzyme product is defective and the pathway is interrupted at a specific step. Unlike the prototrophic parent, the auxotrophic mutant can grow only if the end product of the pathway is available in the growth medium.

Lederberg made two different strains of *E. coli*, each carrying several auxotrophic mutations, mixed them, and spread millions of bacteria on a "minimal" medium lacking the compounds the strains needed to grow, whereupon a few colonies grew. Since neither strain cultured separately yielded any such prototrophs, a rare gene exchange must have occurred in which the good copy of some of the genes had been transferred from one of the two strains to the other and had replaced the damaged copies of those genes. Lederberg had brilliantly cut through all the doubts inherent in previous attempts to see bacteria mating and had proved that *E. coli* used some process equivalent to sex in animals and plants to exchange genes. I set out to repeat Lederberg's experiment with *Streptomyces*.

Lewis Frost had already obtained half a dozen *Streptomyces* cultures from George Floodgate, a friend of his at the Royal Technical College in Glasgow, Scotland (later the University of Strathclyde), so, as I started my project in the Cambridge Botany School, I spread them on agar plates and picked one that not only grew well but also made a beautiful blue pigment, later shown to be an antibiotic called actinorhodin (Color Plate 2B). I had read enough to know that this made it a strain of *Streptomyces coelicolor*: *coelicolor* means "heavenly colored" or "sky colored." (Confusingly, actinorhodin means a red compound made by an actinomycete; the paradox is explained by the litmus-like ability of actinorhodin to be red in the acid conditions used by chemists to study it but blue in the slightly alkaline culture medium.) I began by isolating some auxotrophic mutants that could not grow on the simple medium that supported the wild type unless an amino acid was added—histidine for one mutant, methionine for another. This was easy to do, using an elegant trick that Lederberg and his wife Esther had invented, called replica-plating.

In replica-plating, a piece of sterile velvet covering a circular block that just fits inside a Petri dish is pressed onto the agar surface on which many colonies are growing on a rich medium. The velvet picks up bacteria from each colony, and, when pressed onto a second dish containing minimal medium, it transfers a "print" of the set of colonies. After incubation, colonies of the wild type can grow on the replica, but auxotrophic mutants cannot; the latter therefore can be identified even in the presence of hundreds of unmutated colonies. Figure 3.4 shows the replica-plating

Figure 3.4. Replica-plating tools, as part of the simple equipment needed for microbial genetics in the mid-1950s. In the front, from left to right, are seen a supply of velvet squares for replica plating, a Petri dish containing a *Streptomyces* culture with its lid removed ready to make a replica, and a replicating block with a velvet square held in place by a metal ring. Also seen are a copper can of Petri dishes, two other Petri dish cultures with lids in place, a wooden test-tube rack with slant cultures of *Streptomyces* and an inoculating wire, a Bunsen burner, a beaker of disinfectant for soiled pipettes, and a metal can of clean pipettes. (Courtesy of Susan Bunnewell, John Innes Centre.)

device as part of the simple "place-setting" needed to do microbial genetics in the early 1950s, and Figure 3.5 is an example of the outcome. Colonies thus identified as auxotrophic mutants are tested for growth on a range of substances to see which one they can no longer make for themselves.

I mixed my two different mutant strains, let them grow together until new spores were produced, then spread millions of the spores on minimal medium. A few colonies appeared (Figure 3.6), but none grew when spores from either of the two parental strains cultured separately were spread on the same medium. I had discovered that *Streptomyces* evidently does have "sex." I could now exploit it to find out about how genes control the special features of streptomycetes, and a career-long love affair with these microbes began. Meanwhile, the idea that the actinomycetes were intermediate between bacteria and fungi came under challenge.

Streptomyces Are True Bacteria

Through a chance encounter, I collaborated in a study of the microscopic structure of the actinomycetes that led to a clear understanding of their relationships. Early in my PhD project, I was lucky enough to get to know Audrey Glauert in Cambridge, where most things seem to happen in or around pubs: James Watson and Francis Crick's discussions of the structure of DNA is the most famous example. A mutual friend introduced me to Audrey (Figure 3.7) at lunchtime one day at the beginning of October 1956 on the grass outside The Mill, a well-known watering hole down by the river Cam. She was working on mycobacteria with Ernst Brieger at the Strange-

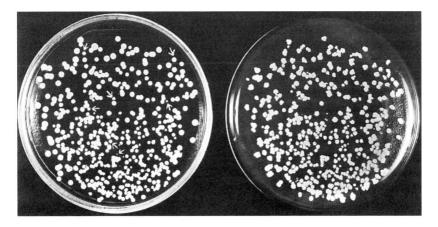

Figure 3.5. Replica plating to isolate auxotrophic mutants of *Streptomyces coelicolor*. On the left is a plate of colonies growing on a nutritionally complex medium spread with a suspension of spores that had been irradiated with ultraviolet light. Arrows mark colonies of auxotrophic mutants that fail to grow on the replica plate on the right, which contains a simple, "minimal" medium. (From Hopwood, D. A. [1959]. *Genetic recombination in Streptomyces coelicolor*. PhD thesis. University of Cambridge, UK.)

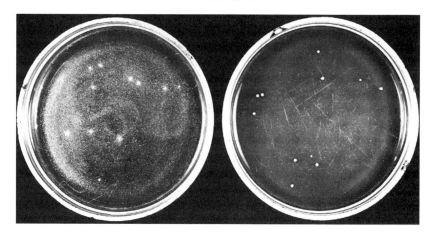

Figure 3.6. Some of the first genetic recombinants of *Streptomyces coelicolor* growing on selective media spread with spores from a mixed growth of two auxotrophic mutants. Out of millions of spores on the plates, just a few colonies—the recombinants—have grown. On the left-hand plate there is a faint background growth from parental spores. (From Hopwood, D. A. [1959]. *Genetic recombination in Streptomyces coelicolor*. PhD thesis. University of Cambridge, UK.)

ways Laboratory on the edge of the Gogs, the slightly rising ground 3 miles South of Cambridge that counts as hills in that part of the world. It had just become possible to cut very thin sections of bacterial cells by embedding colonies in resin and cutting thin slivers on a knife made by fracturing plate glass in a particular way, then examining the sections in the electron microscope.

Audrey's twin brother, Richard, worked as a chemist at Aero Research Ltd. in Duxford outside Cambridge, where epoxy resins, developed initially by the Ciba Company in Basel, Switzerland, were manufactured under the trade name Araldite. They had been invented for gluing airplane components together, as in the Comet, the first commercial jet airliner. None of the standard Araldite formulations was suitable for microscopy, but Richard provided Audrey with components that could be mixed to give a resin with just the right properties for embedding biological specimens. The resin needed to be soluble in the alcohol or acetone used to extract water from the specimens, to have a low viscosity before the hardening process was started so that the resin could penetrate the cells, and to not shrink during hardening and thereby distort delicate biological structures. As a result of Audrey's work, epoxy resins quickly replaced the less satisfactory methacrylate, which had been used for preparing sections of animal tissues for electron microscopy.[6]

I found Audrey's dynamism infectious after the rather laid-back attitude to research in the Botany School. She had trained as a physicist and approached biological research with a refreshing rigor. She was an enthusiast for life and science and did not believe in wasting time. I took the first *Streptomyces* specimens out to the Strangeways Laboratory only a few days after our first meeting, and very soon they were embedded in resin and ready for sectioning. Several times a week, I would cycle

Figure 3.7. Audrey and Richard Glauert on the "Kronprinz Frederik" en route to the First European Regional Conference on Electron Microscopy in Stockholm, September 1956, at which they described the advantages of using Araldite for electron microscopy of bacteria. (Courtesy of Audrey Glauert.)

out and spend happy afternoons with Audrey, peering into the electron microscope at thin sections of *S. coelicolor*. We soon realized we were looking at an organism with a typical bacterial anatomy, like that of bacilli, which others were just beginning to study (Figure 3.8). Crucially, there was no membrane separating the center of the cell, where the DNA was visible as thin filaments, from the rest of the cell, where the dot-like ribosomes were densely packed. In other words, the cells did not have a true nucleus, so the organism was a prokaryote, like a typical bacterium, not a eukaryote like a fungus. Audrey and Ernst Brieger found a similar anatomy in the mycobacteria.

This discovery prompted me to look more critically at earlier publications, and I was soon convinced that the idea of the actinomycetes as a group intermediate between prokaryotes and eukaryotes was probably wrong. The mat (or mycelium) of elongated, branching filaments (hyphae) typical of *Streptomyces* had fooled some of my predecessors into thinking of them as fungi, but it was a superficial resemblance. It seemed clear that this growth form had developed quite independently among the prokaryotic actinomycetes and the eukaryotic fungi, in response to similar selective pressures, after their unicellular progenitors had diverged at a time later estimated to be more than 2 billion years ago.

I read a paper by Cecil Cummins and Harry Harris, at the London Hospital, reporting that the cell walls of *Streptomyces* and other actinomycetes had the same chemical composition as those of bacteria, with chains of special sugar units, one of them unique to bacterial cell walls, cross-linked by short bridges of amino acids to

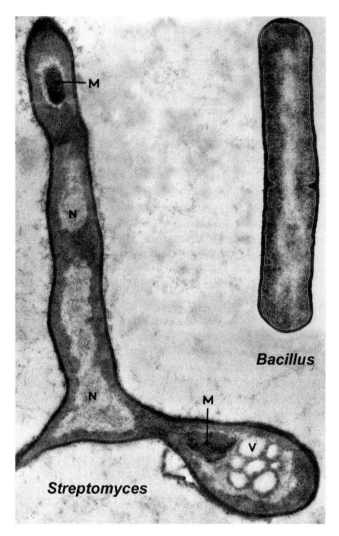

Figure 3.8. Electron microscope images of a thin section of a germinating spore of *Streptomyces coelicolor* and of a dividing *Bacillus subtilis* cell. In both organisms, note the central pale areas containing DNA (labeled N in the *Streptomyces* image), the cytoplasm containing the dot-like ribosomes, and the dense cell wall. M and V in the *Streptomyces* image are intracellular membranes and a vacuole. (This is a montage of two separate images, provided by Audrey Glauert for *Streptomyces* and Jeffrey Errington for *Bacillus*.)

give the equivalent of a corset.[7] In contrast, the walls of fungi are made of long unbranched chains of simple sugars in the form of chitin or cellulose. People were also finding most streptomycetes to be sensitive to antibiotics, such as streptomycin, that killed bacteria but had no effect on fungi. The viruses that attacked streptomycetes provided another clue. They were tadpole-like, with a long tail that attached to the host cell (Figure 3.9), just like the bacteriophages of *E. coli* that were becoming the

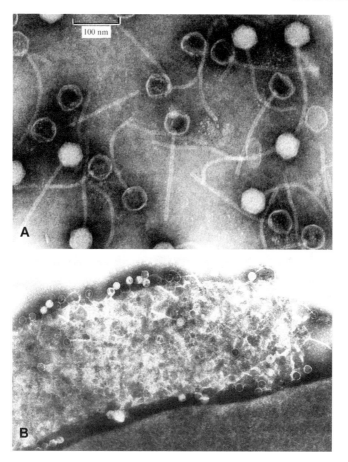

Figure 3.9. Electron micrographs of *Streptomyces* bacteriophages. (A) Purified phage particles (some are intact, with light heads, whereas others have lost their DNA and contain a dense metallic salt); the scale-bar represents 100 nanometers, one ten-thousandth of a millimeter. (B) Phages attached to a degraded *Streptomyces* hypha. (From Dowding, J. [1972]. *Studies on bacteriophages of Streptomyces coelicolor.* PhD thesis. University of East Anglia, Norwich, UK.)

stars of molecular biology. No similar virus attacks eukaryotes. So the actinomycetes were likely to be a group of bacteria, distinct from simple organisms such as *E. coli* and *Bacillus,* but with no special evolutionary relationship to the fungi.

Most microbiologists studying the actinomycetes from the late 1950s on, including Waksman, came to the same conclusion, with just a few hanging on to the idea of actinomycetes as distinct from both bacteria and fungi. The situation was finally settled in the 1970s, after Carl Woese at the University of Illinois developed a powerful new way to deduce evolutionary relationships.[8] Others had already realized that, over time, organisms accumulate random changes in their genes, and therefore in the proteins they encode, as long as they do not impair the proteins' functions. This provides a biological clock to measure the degree to which organisms are related, assuming that the changes accumulate at a constant rate.

Woese reasoned that the sequences of bases along the RNA molecules that form the permanent framework of the ribosomes (not the messenger RNA molecules, which are transiently associated with the ribosomes as they are translated into protein) are ideal biological clocks. This is because their overall architecture is conserved throughout all forms of life, but their precise sequence is free to vary within limits. Woese used his technique to construct phylogenetic trees and found that the bacteria and the eukaryotes formed two distinct branches. *Streptomyces* and the other actinomycetes fitted into the tree of life next to gram-positive bacteria such as *Bacillus* (Figure 3.10). Ironically, Woese did identify a third group of organisms, separate from both the bacteria and the eukaryotes, but these were not the actinomycetes. They were unicellular prokaryotic organisms, many of them adapted to extreme conditions such as hot springs or very acid environments and previously regarded as true bacteria. He called them the Archaea and elevated them to a third kingdom of life.

Since I had taken up a study of the actinomycetes precisely because of their presumed intermediate phylogenetic relationships, the realization that this was a misconception might have been expected to sap my enthusiasm for continuing to work on them, but as far as I remember this did not cross my mind. On the contrary, even if true bacteria, they were evidently only distantly related to the few bacteria already studied from a genetic point of view, and those showed a remarkable diversity of genetic behavior, so the actinomycetes were almost bound to be different again. Meanwhile, I had discovered that I was not alone in this endeavor. As often happens in science, there is a ripe time for an idea, and at least five others had started to look for genetic recombination in *Streptomyces* at about the same time, unbeknownst to each other: Lederberg's graduate student Gaylen Bradley at Stanford, Donald Braendle working with Waclaw Szybalski in the institute at Rutgers that Waksman had built from the streptomycin royalties, and researchers in Moscow, Tokyo, and Rome.

Giuseppe Sermonti and his wife, Isabella Spada-Sermonti (Figure 3.11), in Rome, published the first account of genetic recombination in *Streptomyces* in 1955, so I had a worrying time when Harold Whitehouse tipped me off about the issue of *Nature* magazine in which they announced their results. They had even used a strain of *S. coelicolor* for their work, for the same reason as I had—we thought variations in its beautiful blue color would make good markers to follow gene inheritance. This setback meant that I had to go a bit further than just proving the existence of gene exchange before I could publish, so I pressed on and managed to build a first genetic map charting the positions of a few of the genes on the chromosome. This was not easy, as it would have been in a fungus with a sexual life cycle; it took a couple of nail-biting years as the PhD clock ticked away.

Mapping *Streptomyces* Genes

Fungi usually have a single set of chromosomes—the haploid set—in each cell nucleus for most of the time, maintained by vegetative nuclear division (mitosis). When sexual reproduction occurs, gametes are produced that fuse to yield cells with two sets of

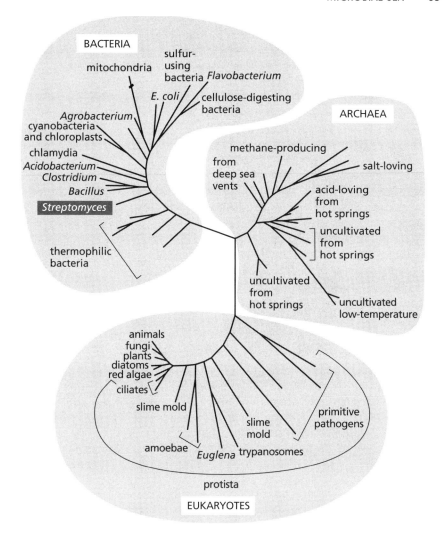

Figure 3.10. The universal tree of life based on ribosomal RNA sequences. (Modified from a version kindly supplied by Susan M. Barns, Los Alamos National Laboratory, New Mexico.)

genes, called diploid. These immediately undergo a special form of nuclear division, meiosis, to yield haploid cells again, but with the chromosomes reassorted. In higher plants and animals, the situation is reversed: they are diploid for most of the time, and meiosis occurs to produce the haploid eggs and sperm, which fuse to give the next generation. However, these differences are only in the relative lengths of the haploid and diploid phases: in all eukaryotes, it is the regular alternation of diploidy and haploidy, separated by meiosis, that is crucial.

Genes on the same chromosome are also reassorted during meiosis. Corresponding chromosomes inherited from the two parents pair with each other, and a special

Figure 3.11. Giuseppe Sermonti and Isabella Spada-Sermonti in 1953 at the Istituto Superiore di Sanità, Rome. (Courtesy of Giuseppe Sermonti.)

process called crossing-over breaks the DNA molecules that run the length of the chromosomes at equivalent points and rejoins the pieces in new combinations. Crossing-over provides the basis for mapping the positions of genes on the chromosomes. Two parents that have different versions of several genes are mated, and the numbers of the various new combinations of genes (new genotypes) that appear among a sample of the progeny are counted. Because crossing-over occurs at random along the chromosomes, the frequencies of new genotypes reflect the genes' relative positions: the more distant the genes, the more often they reassort. Therefore, one ends up with a series of linear maps, like railway lines with stations along them, each line representing one of the chromosomes and each station a gene.

Making genetic maps in bacteria was much more difficult. One problem was that the various bacterial gene exchange processes occurred rarely, so that new combinations of genes had to be selectively recovered, as Lederberg had done in *E. coli* and I had repeated with *S. coelicolor*. This meant that, although some types of progeny with new combinations of genes could grow, others could not. For example, a new combination that included *both* the defective genes of the parents (for histidine *and* methionine synthesis, in my first experiment) could not be identified. One could not easily work out the frequency of crossing-over from this incomplete data set.

Another problem is that in the bacteria already studied, an incomplete set of the genes from the donor enters the recipient, and therefore each gene exchange can give rise to only some of the possible new genotypes. This had not been shown in *Streptomyces* but it had to be regarded as a strong possibility. Despite these difficulties, I devised a method of mapping genes. This came to me like a bolt from the blue one afternoon while I was explaining my work to two German academics from St. John's College in Cambridge, where we were colleagues, as they visited the laboratory to see the microbes I had been enthusing about. As nonscientists—one was a linguist and the other an economist—they were really struck by seeing a novel solution to a scientific problem appear before their very eyes: the Eureka Principle at work! Most scientific advances are not like that, so it helped to propagate a bit of a myth.

The experiments consisted of mixing two parents, each with two defective genes, so that there were four differences between them. Reassorting them would yield 16 combinations of genes in the progeny, of which 2 would be indistinguishable from the parents, produced in their millions by asexual reproduction, so their frequencies were arbitrary. However, it occurred to me that I could estimate the frequencies of a surprising number of the other 14 genotypes if I spread the same number of spores on several different selective media, on each of which a different set of genotypes could grow, and this was enough to make a genetic map. It showed only the relative spacing of the genes, not the absolute distances, because I could not estimate the frequencies of the two classes of nonrecombinant progeny, but it was a start. I had reached this stage by the end of 1958, in time to write my PhD thesis and submit it the following year.

An Italian Job

I had been in Cambridge for 8 years by then and was looking for a break, so I accepted an invitation from Giuseppe Sermonti to spend time in his laboratory. In April 1960, I drove my stock of *S. coelicolor* mutants and recombinants to Rome and started to collaborate with the Sermontis. They had taken up *Streptomyces* genetics for quite a different reason from mine: they wanted to try to harness the power of genetic recombination to improve antibiotic production. They were working at the Istituto Superiore di Sanità (Higher Health Institute). Its predecessor, the Istituto di Sanità Pubblica (Institute of Public Health) had opened in 1934, with financial aid from the Rockefeller Foundation; one of its components was a unit to work on malaria. This disease was rife in the Pontine Marshes south of Rome, and the Foundation had set up a station to combat malaria as early as 1925.[9] In 1941, the name was changed, and in 1948 the institute assumed a bigger role in the control of infectious disease with the opening of various departments, notably an International Centre of Chemical Microbiology headed by Ernst Chain, who had moved to Rome after losing a battle to obtain what he regarded as essential experimental fermentation facilities in the United Kingdom. As well as research laboratories, the institute had a pilot plant for optimizing penicillin fermentations and even production facilities. Chain strongly supported the aim of rationally improving antibiotic production by genetic crossing, but it was Guido Pontecorvo who originated the idea.

The Pontecorvo family was one of several prominent Jewish clans in the Rome ghetto in postmedieval times. Guido's grandfather, Pellegrino, had moved to Pisa in 1890 to set up what became a large and successful textile business, and his sons, including Guido's father Massimo, managed the factories. Guido was born in Pisa in 1907,[10] the eldest of eight brothers and sisters, at least two of whom also went on to make names for themselves. Gillo became a film director who specialized in works with revolutionary subjects, of which *The Battle of Algiers* (1966) is the best known, and Bruno was a distinguished nuclear physicist, noted for his work on the neutrino, who entered the limelight when he defected from the United Kingdom to the Soviet Union in 1950.

Guido first studied diseases of plants, including oleander, on which he published in 1929, and then became an animal breeder at the Agricultural Advisory Service for Tuscany in Florence, working with the local breeds of cattle. In 1938, fascist Italy promulgated the infamous "law for the defence of the race," and Guido was warned that he would lose his job. He went to Edinburgh, Scotland, where the Institute of Animal Genetics was a center for applying genetics to the improvement of farm animals; he intended to move on to Peru, where a post as a cattle breeder would open the following year. He was in Edinburgh when he received a letter from the Tuscan office of the Ministry of Agriculture and Forestry dismissing him under the 1938 law.[11] The Peru job fell through the following year, but by then he had come under the spell of Hermann Muller, one of the American giants of genetics, who happened to be living in the same institute guesthouse. Muller had been in Moscow working with the renowned plant geneticist Nikolai Ivanovich Vavilov, then served briefly in a blood transfusion unit in the Spanish Civil War before coming to Edinburgh. Guido decided to switch fields and stay in Edinburgh to tackle the fundamental problems of gene and chromosome structure that Muller was pioneering, using X-rays to induce chromosome breakage and recombination, and working on fruit flies. (Muller won a Nobel Prize in 1946 for discovering that X-rays cause mutations.) Guido completed his PhD with Muller in just 2 years.

As an Italian, Ponte, as he became known to friends and colleagues everywhere, was interned on the Isle of Man when Italy entered the war in 1940 but was released after 6 months. He could not return to Edinburgh because the east coast of Britain, under threat from a possible German invasion, was a prohibited zone for enemy aliens, so the professor of zoology in Glasgow, Edward Hindle, offered him laboratory space. Ponte started studying the life cycle of body lice in the hope of finding ways to interrupt it, as his contribution to the war effort; he fed the lice by strapping containers to his leg, because they must have regular meals of fresh human blood. He would stay in Glasgow, with a year's break as a temporary lecturer in Edinburgh in 1944 (Figure 3.12), for more than 25 years, building up a renowned department of genetics.

In 1943, Ponte switched from working on lice to using his knowledge of radiation genetics to try to persuade the penicillin fungus to make more of the antibiotic. However, it became government policy to transfer such research to the United States, and soon Milislav Demerec, head of the Cold Spring Harbor Laboratory, working with a British secondee, Eva Sansome, obtained high-yielding radiation-induced mutants. After Sansome returned to Manchester at the end of 1944, Ponte had a long correspondence with her about their experiences working with *Penicillium*,[11] and Demerec offered Ponte a job in October 1946; however, he felt committed to the scientists who were supporting his appointment in Glasgow and decided to stay.

The experience with *Penicillium* had awakened Ponte's interest in the genetics of fungi as suitable organisms for basic genetics research, and he chose *Aspergillus nidulans* as an ideal subject. With it, he went on to make fundamental investigations

Figure 3.12. Guido Pontecorvo examining a Petri dish of *Aspergillus* in Edinburgh, April 1945. (Courtesy of Lisa Pontecorvo.)

on the structure of genes. As others joined the group (notably Alan Roper, his main collaborator), and initially as a side issue to the main research,[11] they found a way of achieving genetic recombination by a process they called the "parasexual cycle." This consisted of a series of uncoordinated events that, taken together, had the same genetic consequences as a sexual cycle. Both cycles occurred in *A. nidulans,* so the results of sporadic genetic recombination in the parasexual cycle could be compared with those of sexual crosses in which recombination occurred regularly at meiosis.

The parasexual cycle in fungi, as Ponte's group defined it, involved the fusion of hyphae of two genetically different strains of *Aspergillus* to yield a mycelium containing a mixture of two kinds of haploid nuclei, a heterokaryon. Then, occasional nuclear fusion resulted in diploid nuclei, which divided as diploids and colonized part of the mycelium. Once in a while, crossing-over occurred between a pair of corresponding chromosomes, resulting in new combinations of genes. Also as a rare event, the mitotic nuclear division went wrong and nuclei with haploid sets of chromosomes arose, with new combinations of chromosomes. Most of the fungi used in industrial processes, including strains of *Aspergillus niger* used to make citric acid, and *Penicillium notatum,* the penicillin producer, lack a sexual cycle. Ponte proposed that the parasexual cycle could be used as a tool in strain improvement by bringing together desirable characteristics from different selected lines, just as in plant and animal breeding. Prompted by the National Research Development Corporation, a British government–owned body that had been set up to exploit inventions and prevent them going abroad, as had happened with the penicillin discovery during World War II, Ponte and Roper took out a patent on the idea in 1952, entitled "Synthesis of strains of micro-organisms."

In May 1952, Ponte wrote to Ernst Chain in Rome: "In the last 18 months we have made considerable progress in a direction that might interest you. We have developed a technique for obtaining genetic recombination in *asexual* filamentous fungi. This permits [us] to tackle the problem of deliberate 'breeding' of strains in species like *Penicillium, Aspergillus* etc. I need hardly stress the possibilities for the production of high yielding strains."[11] At Ponte's suggestion, Chain sent a geneticist to Glasgow to learn the techniques and take them back to Rome. This was Giuseppe Sermonti, who spent 4 months in Glasgow in the winter of 1952–1953, where he discovered a parasexual cycle in the *Penicillium* fungus, and his name was added to the patent. It was a logical next step for Sermonti to try to do something similar in *Streptomyces*. This led to his 1955 publication and my collaboration with him.

I had two wonderful stays in Rome, in the spring and summer of 1960 and again in the spring of 1961: this was the era of *La Dolce Vita*, as portrayed in Federico Fellini's 1960 movie. Apart from working in the laboratory, I got to know Rome and its surroundings in a way that few tourists can. I spent a lot of time with Alberto Mancinelli, one of the young postdoctoral fellows in the laboratory, his sister Marisa, and his friends. They introduced me to the best pizzerias and coffee bars in the city, and at weekends we would make trips to the seaside or to the mountains in Alberto's tiny Fiat Cinquecento and my Morris Minor, swimming, picnicking, and enjoying the outdoors. A surprising number of public holidays and saints' days were observed, and they provided even more breaks from laboratory routine. The working day was from 9 AM to 2 PM, with a second (somewhat optional) session from about 4:30 until 8 PM. On weekday afternoons, I usually went to one of the hundreds of interesting spots in the city—a Roman relic, an early Christian church, or a quiet park—where Marisa would often help with my do-it-yourself Italian course and generally make sure my knowledge of the language progressed.

In spite of living life to the full, I did some useful experiments and we made progress in understanding *Streptomyces* genetics. I mostly studied a special kind of colony that the Sermontis had just found among the progeny of crosses arising on the media that selected prototrophic recombinants. Instead of breeding true, like most of the recombinants, these colonies gave rise to a mixture of different genotypes. We deduced that the spores they grew from contained a complete chromosome from one of the parents and a fragment from the other. This was important, because it seemed to provide evidence that, as in the bacteria already studied, chromosome transfer probably resulted in incompletely diploid cells. These colonies also offered a practical benefit, because their progeny could be analyzed without having to impose a selection for recombinants (a novel situation in bacterial genetics), making gene mapping more straightforward. We also spent a lot of time evaluating what we knew by then of the processes underlying genetic recombination in *Streptomyces* and what they had in common with those in other bacteria and in fungi.

Gene Exchange in Bacteria

There were the three bacterial precedents to consider before deciding whether *Streptomyces* used a totally novel mechanism to exchange genes.[12] Transformation was discovered by Fred Griffith in London in 1928 in the pneumococcus, and it was shown to involve naked DNA by Oswald Avery and his group at the Rockefeller Institute in New York in 1944. One member of the population liberates DNA, probably by a chance breakup of some of the cells, and a recipient bacterium takes it up. This DNA uptake is a specialized mechanism (now known to involve about 20 genes) that develops only at a certain stage of the growth cycle, as a group response. Those members of the population that first reach this critical stage liberate a special signaling molecule (a very small protein called a peptide) that causes their neighbors rapidly to become "competent" to take up DNA. This is adaptive, because genetic recombination works only when DNA is exchanged between individuals with different genetic make-ups, and the chances of this occurring are increased if a large population engages in gene exchange at the same time. The DNA introduced into the recipient by transformation can replace a corresponding segment of the new host's chromosome by the same process of crossing-over that occurs between whole chromosomes during meiosis in eukaryotes, generating new genotypes. Transformation was later found to occur in *Haemophilus influenzae*, an important respiratory pathogen, and in the soil-living *Bacillus subtilis*; it is now known to be widespread but sporadic among bacteria.

A different process, transduction, occurs as a result of an accident of the life cycle of a bacterial virus. When a bacteriophage attaches to a sensitive bacterium, it injects its DNA into the host, where there are two possible outcomes. Some phages, called virulent, immediately begin to replicate their DNA to makes tens or even hundreds of copies, and the bacterial chromosome is broken into fragments. The phage DNA then directs the dying host cell to produce viral proteins that assemble into new virus particles, each containing a copy of the bacteriophage DNA, and these are liberated from the dead host. Another phage type, called temperate, has a choice of two options, which is exercised in each infection. One is to kill the cell, like a virulent phage, and the other is to insert the phage DNA into the host chromosome by crossing over between specific sequences (attachment sites) on the phage DNA and on the chromosome. Almost all the virus genes are switched off, and the phage DNA (called a prophage) is inherited indefinitely as part of the host chromosome. Then, once in a while, the phage DNA separates from the chromosome, and a round of virus replication and maturation takes place. Stresses such as heat or ultraviolet light can trigger this event. It is an adaptation for the phage's survival, because it is a signal to "leave the sinking ship" in case the stress proves fatal to the host bacterium.

Norton Zinder, working with Lederberg in 1952, discovered the first example of transduction in *S. typhimurium*, a cousin of *E. coli*. This is an important cause of bacterial food poisoning, killing a million people worldwide every year. Edwina Currie famously brought it to the attention of the public in the United Kingdom when she was Health Secretary in Margaret Thatcher's government. Her exaggerated comment in 1988 that all British eggs were infected with *Salmonella* halved egg consumption overnight and led to her resignation.

Zinder had set out to determine whether *Salmonella* exchanged genes in the same way as *E. coli* and discovered that, unlike in that bacterium, he could obtain recombinants using culture fluids from one strain added to another. However, the process differed from transformation by naked DNA: a temperate bacteriophage called P22 was responsible. Transduction happens when, as a rare accident during phage replication—one time in tens of thousands of replications—a fragment of the host chromosome, instead of phage DNA, is packaged by the viral proteins, becoming a dummy bacteriophage. Going on to infect a new bacterium, it delivers the piece of bacterial chromosome from the previous host, which undergoes crossing-over with the recipient chromosome and thus gives rise to recombinant progeny.

Another famous temperate phage is called lambda. It was discovered in Lederberg's laboratory in 1951 as a prophage in the *E. coli* chromosome and was later found to transduce host genes by a different mechanism from P22. Occasionally the prophage exits the host chromosome inaccurately, leaving some phage genes behind and picking up nearby bacterial genes to replace them. Like transformation, transduction was later found in many bacterial groups.

From our studies of genetic recombination in *S. coelicolor*, it was clear that we were not dealing with transformation or transduction. One pointer was that donor genes far apart on the genetic map were often inherited together. This cannot happen in transformation or transduction, because those processes can carry two genes into a new host only if they lie on the same short stretch of DNA. We also found that *Streptomyces* genes could not be transferred by means of a culture medium from which the cells had been filtered off. These results suggested that some kind of mating between bacteria was involved. The only precedent for bacterial conjugation was the system discovered by Lederberg and Tatum in *E. coli*.

Through the work of Bill Hayes at the London Postgraduate Medical School, and in parallel by Luca Cavalli-Sforza in the Lederbergs' laboratory at Stanford, this process had been found to differ from any mating known in eukaryotes. The key to elucidating the sexual system of *E. coli* was the realization around 1952 that various descendants of Lederberg's original strain differed in their ability to exchange genes. The starting strain was postulated to contain a genetic element separate from the chromosome, which was called the F (for fertility) factor. It caused the cells to mate, producing channels through which DNA could pass. Lederberg invented the name "plasmid" to describe such freestanding genetic elements.

In the early 1960s in Japan, in work pioneered by T. Watanabe, and very soon worldwide, other gram-negative bacteria were found to harbor sex plasmids that were like F but carried antibiotic resistance genes, which were transferred to a new host during mating. This seminal discovery explained how resistance genes spread rapidly within a population of bacteria. Other plasmids carry genes that allow bacteria to grow on a nutrient they cannot otherwise use, such as the hydrocarbons that make up petroleum. Bacteria carrying such plasmids can be used to clean up oil spills. It was a U.S. patent filed in the name of Anand Chakrabarty in 1971 that pioneered the patenting of artificially created microbes, which had hitherto been disallowed: it described the construction of a bacterium carrying several such plasmids, enabling it to break down each of the major hydrocarbons in crude oil. Plasmids that are important in infectious disease carry genes that help the bacteria to infect an animal or plant host for which they would not otherwise be pathogenic. Many cases of bacterial gastroenteritis are caused by such plasmid-carrying strains of *E. coli*.

Among Lederberg's collection of *E. coli* strains, some had lost the F plasmid, and in others it had become integrated into the chromosome. A bacterium carrying the F plasmid in its original state (F$^+$) behaved like a male and could mate with a plasmid-free recipient, transferring a copy of the plasmid into the F$^-$ cell. When both parents had the plasmid it was very rarely transferred, and then probably only when an individual in one of the cultures lost the plasmid and so could behave as a female. When a strain with the plasmid integrated into the chromosome (called Hfr for "high frequency of recombination") was mixed with an F$^-$ culture, every time the plasmid moved into a female it dragged the chromosome with it, and many recombinants appeared. In contrast, F$^+$ × F$^-$ matings gave few recombinants, presumably resulting from occasional plasmid integration into the chromosome to produce an Hfr bacterium in the F$^+$ population.

Usually, the union between the two *E. coli* cells breaks during the mating, so an Hfr bacterium transfers only a segment, on average about one fifth, of the donor chromosome to the recipient. The union can be broken prematurely, by violent shaking of the culture, as was first shown by François Jacob and Elie Wollman at the Pasteur Institute in Paris. This procedure became the standard means of mapping the genes on the chromosome, using the time at which they enter the recipient as a scale. To do this, Hfr and F$^-$ cultures are mixed, and samples of the mating bacteria are taken at different times and shaken to break the connections between them. The culture is then spread on media that select for the growth of progeny that have inherited particular genes from the donor parent. Donor genes near to and on one side of the point of integration of the F plasmid are found in progeny selected at early times, whereas genes farther along the chromosome can be found only in progeny selected later. It takes about 90 minutes for the whole chromosome to be transferred if mating is not interrupted. The French researchers also found that different Hfr strains donated genes in a different order, each a "circular permutation' of the others. This allowed them to make the remarkable deduction that the *E. coli* chromosome was a circle into which the F factor could integrate at various positions, resulting in Hfr strains that donated genes starting at different points.

As became clear later, both the F plasmid and the bacterial chromosome are circular DNA molecules, and when they join to give the Hfr state they form a single circle (Figure 3.13). The joining occurs through copies of a special sequence of DNA, called IS (for insertion sequence), that is present at several places on the plasmid and chromosome. These IS sequences provide localized regions in which the DNA of the plasmid and chromosome exactly match, so crossing-over can occur between them. In a mating with an F$^-$ bacterium, the first DNA to move into the recipient is part of the plasmid, and host DNA follows on. After a copy of the whole chromosome has been spooled into the recipient, the rest of the plasmid brings up the rear, if transfer has not been interrupted.

While we were collaborating in Rome, the Sermontis and I found that various mutant and recombinant *S. coelicolor* strains gave very different frequencies of recombinant progeny when mated with each other, and it seemed most likely that this reflected the involvement of a plasmid fertility factor, as in *E. coli*. But we did not

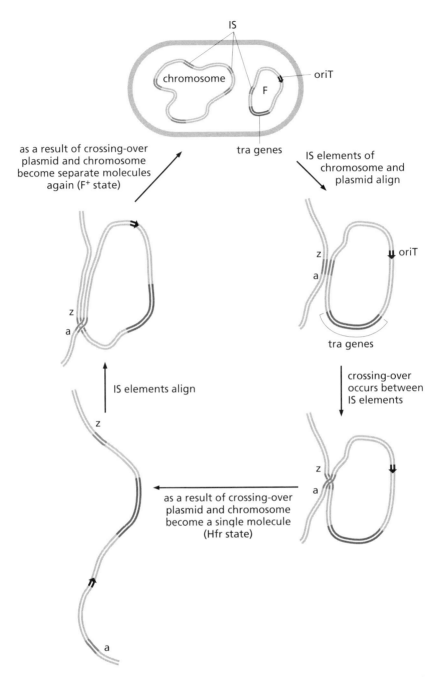

Figure 3.13. Integration of the F plasmid into the *Escherichia coli* chromosome. At the top is an F⁺ bacterium with the F plasmid separate from the chromosome. F carries an IS sequence, transfer (tra) genes and the sequence (oriT) involved in DNA transfer in a mating. Recombination between copies of the IS element on the plasmid and on the chromosome joins the two circular DNA molecules, to produce the Hfr state. Recombination between the two IS copies on the Hfr chromosome reverses the process of integration. a and z are bacterial genes on either side of the IS on the F⁺ chromosome; they are separated by the integrated F plasmid on the Hfr chromosome and so are transferred early and late, respectively, if the Hfr bacterium mates with an F⁻ bacterium.

know whether the mating mechanism determined by the hypothetical plasmid resembled that in *E. coli*, and it was only much later that the *Streptomyces* system began to be clarified and turned out to be very different (see Chapter 6).

During my second visit to Rome in 1961, Sermonti and I wrote a long review taking stock of what we had learned about *Streptomyces* genetics and highlighting unanswered questions concerning the mating process and the topology of the chromosome or chromosomes. Sermonti was in a melancholic phase and suffered from insomnia, so he would write sections of the manuscript in wakeful periods and I would translate them into English the next day. It seemed like a natural break point for me scientifically. While I was still doing my PhD, I had been appointed to a temporary position in the Cambridge Botany School, and the 5-year contract was almost up, so I needed a new job. Just then, a faculty post was advertised in Ponte's department in Glasgow, and I applied. He appointed an internal candidate, Bernie Cohen, but he persuaded the Principal of the university to create a 1-year position for me. As I discovered later, he had a deep suspicion of people trained in Oxford and Cambridge and wanted personal knowledge of any candidate before committing to a permanent job. I thought it was worth the risk, so I packed my bags and drove north. It turned out to be a great move, both professionally and personally. Not only did the job mysteriously become permanent, but within a couple of months Bernie had introduced me to Joyce Bloom, a mouse geneticist he had known as a colleague in a previous position at the Institute of Animal Genetics in Edinburgh. We were married in September 1962.

Genetics in Glasgow

I found Ponte an enormously stimulating influence both in and out of the laboratory. He was an iconoclast who had no time for what he saw as petty concerns. He was a constant thorn in the side of the Glasgow university administration, running the department with a half-time secretary and a wastepaper basket (mainly the latter). Apart from his burning curiosity about genetics, he was passionately interested in alpine plants and hill walking and would lead wonderful Sunday outings to the mountains, which start only a half-hour drive out of Glasgow (Figure 3.14), his small, slightly stooping, strongly tanned figure striding out ahead. When Ponte died in 1999, just short of his 92nd birthday, it was as a result of a fall while on an Alpine mushrooming walk.

By the time I joined Ponte's department (Figure 3.15) in 1961, he was pioneering human cell genetics. He aimed to discover a kind of parasexual cycle in cultured cells and use it to study human heredity without the constraints imposed by people's sexual choices. Work on *Aspergillus* was still going on, but the focus of the department had shifted away from microbes. Ponte asked me, "Are you going to carry on with *Streptomyces*?" I replied rather tentatively, "For the time being," to which he countered rhetorically, "Why not forever?" I took this as a sign of approval, still not feeling committed for all time. In the end, there never seemed to be a good reason to switch. However, there was no suggestion that I should work on antibiotics, because Ponte had become disillusioned at the lack of interest on

Figure 3.14. Guido Pontecorvo and members of the Glasgow University Genetics Department on an outing to Loch Lomond, June 26, 1954. Left-hand group (from left to right): Guido Pontecorvo, Etta Käfer (back to camera), David Wilkie, Robert Pritchard; central group (from left to right): Elaine McKendrick, Margaret Forbes (back to camera), Neal Cleat, Ted Forbes, unidentified; right-hand side: Alan Roper. (Courtesy of Lisa Pontecorvo.)

the part of the drug companies in applying genetics to improving their antibiotic-producing cultures. He was still cynical about this in 1974 when, in a Presidential Address at the second in a series of international meetings on the genetics of industrial microorganisms—the GIM series that began in 1970 (as described in Chapter 4)—he provocatively stated: "The main technique used [in strain improvement] is still a prehistoric one: mutation and selection.... Nature, in its remarkable ways of coping with the improvement of living organisms—what we call evolution—has given up the exclusive use of mutation and selection at least one billion years ago. It has supplemented mutation and selection with a wonderful variety of mechanisms for the transfer of genetic information."[13]

After I moved to Glasgow, I kept up a regular correspondence with Sermonti for a year or two. He took up the chair of Genetics at the University of Camerino in 1964, later moving to Palermo and eventually Perugia, but he gradually moved away from mainstream science, taking an increasing interest in philosophy and literature. (Luigi Pirandello, the Sicilian playwright who explored the contrast between reality and fantasy, was a cousin of Sermonti's maternal grandmother, and Sermonti's grandfather had been Pirandello's friend and lawyer, so Sermonti may have felt an affinity there.) By the early 1970s, he was arguing against science, or at least scientific reductionism, as the preferred way of interpreting the natural world—he published *The Twilight of Scientism* in 1971—and today he is Italy's

Figure 3.15. The genetics department of the University of Glasgow in 1984. This 10-story building on Dumbarton Road was opened in 1966 and later named the Pontecorvo Building. Previously the department had space in the old Anderson College of Medicine, the Victorian building in the foreground, where David Livingstone studied medicine: a cross-section of a tree trunk in the lobby was reputed to be from the tree where Henry Stanley met Livingstone on the shores of Lake Tanganyika in 1871, with the words: "Dr. Livingstone, I presume?" (Courtesy of Bernard Cohen, University of Glasgow.)

best-known anti-Darwinist (his 1999 book was called *Forget Darwin*), welcomed by supporters of the newest manifestation of creationism, intelligent design. The other groups who had demonstrated genetic exchange in various *Streptomyces* species had found it tough going and switched to other research topics, so I was plowing a bit of a solitary furrow. I pressed on to try to develop the genetics of *Streptomyces* to the point where it could be used as a tool in understanding the special features of their biology.

At first, the kinds of genetic analysis Sermonti and I had been doing revealed two linear maps of genes (Figure 3.16). Maybe *S. coelicolor* had two chromosomes? I didn't think so, because otherwise, without a nuclear membrane and the mitotic apparatus that eukaryotes use to ensure that daughter sets of chromosomes are accurately distributed at cell division, how would this happen? Sure enough, by considering in more detail the pattern of inheritance of the genes in one of the groups in relation to those in the other, I closed the gap between the two original maps by 1965—in fact, I closed both gaps, coming up with a circular map. This was not a total surprise either, in light of Jacob and Wollman's deduction that the genetic map of *E. coli* was circular. Perhaps chromosome circularity was the hallmark of a bacterium, whereas all eukaryotes have linear chromosomes. However, physical evidence is always needed to prove chromosome circularity.

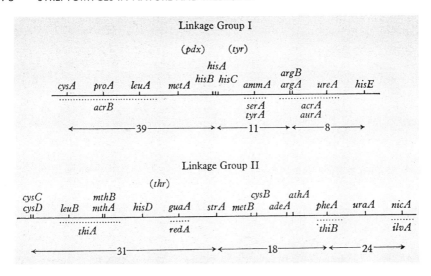

Figure 3.16. *Streptomyces coelicolor* genetic map of early 1965 showing two groups of genes, mostly recognized by auxotrophic mutants. This is the last version of the map before the two linkage groups were joined into a circular map. The *redA* gene was later shown to be part of the actinorhodin biosynthetic gene cluster and is the first antibiotic gene to be mapped.

John Cairns obtained such evidence for *E. coli* in 1963 when he managed to visualize in the light microscope the entire bacterial chromosome spread out as a closed loop on a microscope slide,[14] but no one else could get the technique to work. Frank Stahl, a brilliant American molecular biologist working on bacteriophage genetics, had pointed out that circularity of a genetic map could arise automatically, just as a consequence of the incomplete nature of the mating process, even if the chromosome were linear.[15] The reason is explained in Figure 3.17. I could not solve the ambiguity about whether the *Streptomyces* chromosome was linear or circular with the tools available, but meanwhile I started to use the circular genetic map of *S. coelicolor* (Figure 3.18) as a basis for studying one of the specialties of *Streptomyces*, its sporulating life cycle, without worrying too much about the underlying architecture of the chromosome. Another career move provided the opportunity to study *Streptomyces* genetics in much more depth.

Time to Move On

By 1968, Ponte had had enough of running a university department and decided to move to a nonadministrative position at the Imperial Cancer Research Fund laboratories in London. My wife and I were also ready to move back to England with our three young children. Just then, a unique job opportunity opened up at the John

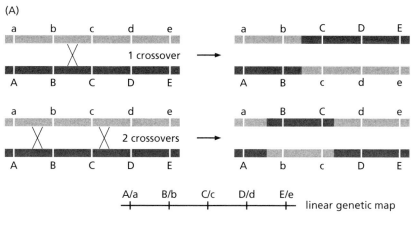
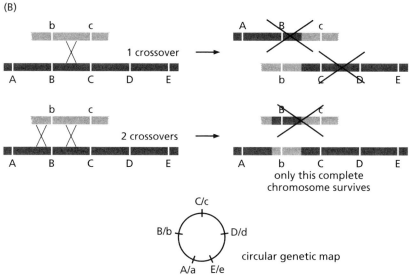

Figure 3.17. Consequences of crossing-over between complete and incomplete chromosomes. (A) When both chromosomes are complete, single as well as double crossovers give full-length recombinant chromosomes and hence viable progeny, and the genetic map is linear. (B) When one of the parental chromosomes is incomplete, only double crossovers give the full-length chromosomes needed for viable progeny, and in these both ends come from the same parent, so genes near the ends of the chromosome are almost always inherited together, as if they were close on the chromosome. This makes the two ends of the chromosome appear to be adjacent on the genetic map, making the map circular.

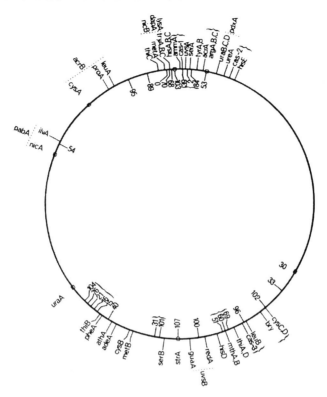

Figure 3.18. *Streptomyces coelicolor* genetic map of 1966 with the two groups of genes in Figure 3.16 joined in a circle. Numbers on the inside of the circle are the positions of a class of mutations that prevent the organism from growing at a temperature of 37°C, while still growing at 30°C; they identify essential genes of unknown function.

Innes Institute in Norwich, and I was lucky enough to be appointed. Thus began a new phase of my career, in which the natural process of gene exchange was supplemented by more powerful artificial procedures. This was a prelude not only to learning much more about chromosome transfer and the structure of the *Streptomyces* chromosome, but also to the practical goal of genetically engineering the production of novel antibiotics.

4

Toward Gene Cloning

My career move in 1968 came at a good moment. Norwich turned out to be an excellent place to bring up a family, and, after a rocky start, I was able to build up a flourishing research group. We continued to exploit the natural mating system of *S. coelicolor* as a tool for studying its genetics. Sex plasmids were discovered, and genetic mapping was refined. Morphological mutants revealed genetic switches controlling progression through the various stages of colony development, and mutants unable to produce the antibiotic actinorhodin, which gives the organism its beautiful blue color, served to identify a cluster of genes for its biosynthesis. *Streptomyces* genetics gained visibility as a worldwide community of academic and company researchers took up the study, some driven by curiosity, as the best academic research usually is, and others to underpin the rational improvement of industrial antibiotic production.

Meanwhile, in the mid-1970s, the science of genetics underwent a revolution with the invention of "recombinant DNA" technology at Stanford and the University of California at San Francisco by Paul Berg, Herb Boyer, Stan Cohen, and their colleagues. New combinations of genes could be made in the test tube and introduced into *E. coli*, yielding a "clone" of genetically modified bacteria. Soon, mammalian genes were expressed, resulting in bacterial cultures that can make human insulin or growth hormone in a fermenter, and a biotech industry was born. Fundamental knowledge about molecular biology came thick and fast too, as the new techniques were used for academic studies. This chapter describes how we applied these advances to *Streptomyces* in order to use the power of recombinant DNA to illuminate its biology.

Genetics in Norwich

In 1967, there was a significant event for British biology. An old, established plant research institute moved to Norwich and affiliated with one of the new "plate-glass" universities that had recently been established by the U.K. government to widen access to university education. The institute was the John Innes Institute (JII) and the university the University of East Anglia (UEA).

The JII had been founded through a bequest to open an institute for "scientific horticulture" by John Innes, a London merchant with an amateur interest in gardening. The John Innes Horticultural Institution, as it was first known, opened in 1910 in his former house at Merton, near Wimbledon in southwest London, with William Bateson as its first director. Bateson was a founder figure in British genetics.[1] He was one of the first to realize the importance of Mendel's mid-19th century work on peas, which laid the foundations of the science of genetics when it was "rediscovered" in 1900, and he directed much of the work of the institute towards plant genetics right from its inception. The JII established an excellent reputation through the 1920s and 1930s. It pioneered studies on many aspects of genetics, including the inheritance of flower pigments, one of the precursors of Beadle and Tatum's work on gene action in *Neurospora*. Much of the basic behavior of chromosomes, later applied to understanding human conditions such as Down syndrome, was also worked out, using the large, easily studied chromosomes of plants as model systems. In 1949, the institute needed to expand, and the director, Cyril Darlington, persuaded the trustees to move the JII to an 18th-century estate, complete with mansion, at Bayfordbury in Hertfordshire, some 25 miles north of London. Probably he fancied the idea of being lord of the manor there, and the trustees fell for his argument that the soil was perfect for growing experimental crops, which it turned out not to be, but for a while the JII flourished. However, after Darlington's departure for the chair of botany at Oxford in 1953, the JII became increasingly isolated, with falling scientific standards. By the mid-1960s, it was time to move again, and this time an idea that had been pre-empted by the move to Bayfordbury was resurrected: the institute moved to a university setting in the hope that association with an academic institution would revive the JII's fortunes.

UEA was one of the rising new universities, with imaginative programs in biology. It was among the first British universities to abandon the old subdivisions of the subject, such as botany, zoology, and microbiology, in favor of an integrated school of biology with interests ranging from biophysics to ecology. In the period of uncertainty before the JII moved to Norwich, several of the more ambitious scientists left, so it was a rump of an institute that relocated in temporary accommodation on a green field site just off the edge of the UEA campus. The new director was Roy Markham, a plant virologist who had headed a research unit in Cambridge. Three chairs, paid for by the JII trustees, had been established; the incumbents were to teach at UEA and run research departments in the JII. I was appointed to the John Innes Chair of Genetics, while Roy Davies, who had worked on the effects of radiation on photosynthetic algae at the UK's Atomic Energy Research Establishment, became John Innes Professor of Applied Genetics, and Roy Markham was John Innes Professor of Cell Biology. He had a low opinion of universities—even in Cambridge he had emphasized the superior research role of institutes and units such as his own over the Botany School. He discouraged Roy Davies and me from playing a significant role at UEA, but we persisted in doing so, and an excellent symbiotic relationship between the two institutions developed over subsequent decades. Researchers at the JII could sharpen their wits by contact with argumentative students, while the students benefited from being exposed to what became a high-powered research environment.

I arrived in Norwich in September 1968, accompanied by the four members of my group from Glasgow, including Alan Vivian, who had been a postdoctoral fellow with me for a couple of years, and Helen Ferguson (later Helen Wright and eventually Helen Kieser), who had joined me in 1965 at age 19 and was to stay in my group until I retired in 1998. We had to build microbial genetics almost from scratch in an environment used to working on plants, but pretty soon the laboratory was up and running. Administratively, all was not plain sailing, however.

When I happened to come across the minutes of an earlier meeting of the governing body of UEA, I learned that the Agricultural Research Council (ARC), who picked up the tab for current expenditure in the institute (the John Innes Trustees paid for the land and buildings), had not given an ongoing commitment to fund the Genetics Department at the JII after its move to Norwich. It might be detached to join biological sciences at UEA. It also transpired that there was little understanding of microorganisms on the governing council of the JII, which had been concerned mainly with plant genetics and breeding, in line with John Innes's original bequest, although from time to time ground-breaking research on fungal genetics had been carried out. I managed to redress the balance a little by having Ponte appointed to a vacant slot on the council to give me moral support, but a first site visit on behalf of the ARC nevertheless gave me the thumbs-down for wanting to continue to work with *Streptomyces* and for a general lack of vision. I asked that Bill Hayes, one of the giants of *E. coli* genetics (Chapter 3), be invited to write a letter pointing out the relevance of my work to genetics in general and so to biology in a broader sense, and he gave me crucial backing, but it was a struggle to get permission to make any new appointments.

The future of the department was gradually assured as I hired new recruits in three areas on the interface between microbiology and plant science, to capitalize on the power of microbial genetics combined with the facilities for plant research at the JII, and to assuage the ARC's fears that they would be paying only for a topic—*Streptomyces* genetics—that they regarded as irrelevant to agriculture. The new projects were to develop a genetic system for *Rhizobium*, the symbiotic bacterium that fixes atmospheric nitrogen in nodules on the roots of leguminous plants; to study a bacterium called *Agrobacterium tumefaciens,* which forms tumors on plants in an intimate relationship involving, as it later transpired through the work of others, the transfer of plasmid DNA into the plant chromosomes by a process of "natural genetic engineering" that became a main route to making genetically modified crops; and to work with wall-less bacteria called mycoplasmas that can live only inside plant cells.

Meanwhile, the *Streptomyces* group expanded and gradually established its credentials, even with the ARC, or at least with its successors (the ARC became the AFRC in 1984 as food was added to its mission, and in 1994 the BBSRC, for Biotechnology and Biological Sciences Research Council). As the other JII departments, working on plants and their viruses, also flourished, the environment for studying molecular biology became better and better. New permanent buildings went up (Figure 4.1), colleagues joined the group, and an ever-increasing army of PhD students and postdoctoral fellows from around the world passed through, enlivening the laboratory as they learned from the permanent members and each other, and we from them. It was an exciting time to be a microbial geneticist.

The most pressing challenge for the *Streptomyces* group as we started in Norwich was to try to make sense of the involvement of plasmids in gene exchange in *S. coelicolor*. While still in Glasgow, Alan Vivian had set about trying to understand the differences that Sermonti and I had observed in the fertility of various recombinant strains derived from my original crosses, and he continued after the move to Norwich. We could not carry out the sort of interrupted mating experiments that had been so informative for *E. coli*, because recombinants arose only if two parental strains were allowed to grow together in a tangled mixture on a solid surface for a few days, not if samples of pregrown mycelium of two strains were put in contact for a while and then shaken apart. However, by observing the inheritance of different fertility levels, and especially finding that differences in mating behavior were transmitted between strains at a much higher frequency than chromosomal genes, Alan firmly implicated a plasmid in this behavior.[2] He called it SCP1, and soon we had examples of strains with and without the plasmid, as well as strains with SCP1 integrated into the chromosome. As in *E. coli*, crosses between two strains lacking the plasmid were the least fertile—

Figure 4.1. The John Innes Institute in 1985. The L-shaped building, which opened in 1973, represented the second phase of permanent building after the institute moved to Norwich in 1967, initially being housed in temporary buildings. The Genetics Department occupied the lower of the two white-clad floors. The pier bearing the name was demolished in a later period of expansion. (Courtesy of John Innes Archive.)

although not completely sterile as in crosses of two F⁻ *E. coli* strains—and crosses between bacteria with and without SCP1 yielded more recombinants. Crosses between a strain with SCP1 in the chromosome and one lacking the plasmid were spectacularly fertile, allowing the isolation of recombinant progeny without selecting against the parents. Janet (Jan) Westpheling, an American PhD student who joined the laboratory in 1976, made valiant attempts to isolate the plasmid. We presumed it was a circular DNA molecule, like all other plasmids known at the time, and therefore should be isolable using published methods suitable for circular plasmids. Jan tried them all, as well as others she invented, but without success. It was a frustrating time for Jan, but her vivacious personality and optimistic approach to life carried her through; she has gone on to make important discoveries in *Streptomyces* molecular genetics (Chapters 6 and 7). Only some 10 years later, through the work of Haruyasu Kinashi at the Mitsubishi-Kasei Institute for Life Sciences in Tokyo and then at Hiroshima University, did the reason for Jan's frustration emerge: SCP1 is a large *linear* DNA molecule that could be separated from chromosomal DNA only by methods (Chapter 5) not yet invented when Jan was in the laboratory.

Meanwhile, Mervyn Bibb graduated from UEA and joined the group as a PhD student. He quickly showed that the low level of fertility in crosses of strains lacking SCP1 was the work of a second plasmid, which he called SCP2. This turned out to be a conventional circle of DNA (Figure 4.2) that could be isolated and visualized in the electron microscope by standard methods, which Hilgund Schrempf applied to *S. coelicolor* during a sabbatical visit to Norwich from the University of Würzburg. This was a big step forward, not only in understanding sex in *Streptomyces*, but, more importantly, as a prerequisite for developing gene cloning. To achieve this objective, we needed to find ways of introducing DNA into the cells. This developed as an

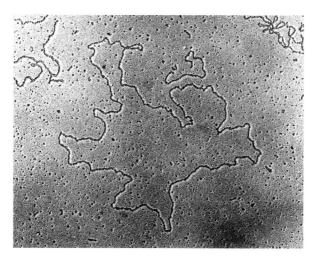

Figure 4.2. Electron micrograph of the circular *Streptomyces coelicolor* plasmid SCP2 (with some broken DNA fragments). Inside the host the DNA would be tightly coiled, because the circumference of the plasmid is about 20 times the diameter of the hyphae. (Courtesy of Mervyn Bibb, John Innes Centre.)

extension of using protoplast fusion as a route to artificial genetic recombination. What are protoplasts and why are they important?

Fusing Bacterial Protoplasts

In 1976, Katalin Fodor and Lajos Alföldi, at the Hungarian Academy of Sciences Institute of Biological Sciences in Szeged, and Pierre Schaeffer, Brigitte Cami, and Rollin Hotchkiss, at the University of Paris in Orsay, announced a new method of recombining bacterial genes.[3] This was protoplast fusion. They had treated *Bacillus* cells with an enzyme called lysozyme, which is found in egg white and tears, where it helps to kill infecting bacteria by destroying their cell walls. (Lysozyme was another discovery of Alexander Fleming, reflecting his long-term interest in natural antibacterial agents.) Lysozyme breaks the cross-links in the bacterial cell wall and allows the contents, still surrounded by the cell membrane, to escape as spherical bodies called protoplasts. The pressure inside a bacterium reaches several atmospheres, and normally this is balanced by the rigid wall, which acts like a bicycle tire, preventing the inflated inner tube from bursting. Protoplasts would explode if the wall were removed in water or in a dilute saline solution, because they would take up water from the medium by osmosis. The osmotic pressure of 10% sucrose in the medium supplies the appropriate stabilizing force.

The *Bacillus* researchers had mixed protoplasts prepared from two genetically different strains and then added polyethylene glycol (PEG), which was already known to cause fusion of animal cells and plant protoplasts. They had then spread the mixture on selective medium, whereupon recombinants appeared. We decided to try this approach in *S. coelicolor*. In order to prepare the protoplasts, we built on earlier studies, especially those of Masanori Okanishi and his colleagues at the National Institute of Health in Tokyo a few years before. They had discovered how to prepare protoplasts of several *Streptomyces* species. More importantly, they had made a painstaking study of the components needed in a special medium to allow the protoplasts to resynthesize the cell wall and become a mycelium again.[4] We soon found that Okanishi's methods were successful for our *S. coelicolor*. Figure 4.3 shows light and electron micrographs of *S. coelicolor* protoplasts, compared with the appearance of the normal mycelium. We could then set about trying to fuse *Streptomyces* protoplasts, which had not previously been reported.

Helen was about to make the first attempt, and it seemed obvious to try the protocol that had proved successful in *Bacillus*. Just then, Ponte visited Norwich for a meeting of the John Innes Council. In a spare hour before the meeting on October 20, 1976, I told him about the planned experiments, and he said: "Forget it, use the conditions that work for animal cells—they will be much better than the ones the *Bacillus* people used." At that time, attempts to bring about gene exchange in human cells in tissue culture ("somatic cell genetics") were in their infancy, and Ponte was one of those trying to improve them. One of the techniques was to fuse human and mouse cells. The products of the fusion had a complete set of chromosomes from each species, but as the cells grew and divided, they gradually lost chromosomes at random. When two or more human biochemical traits were repeatedly lost together, they could be deduced to be

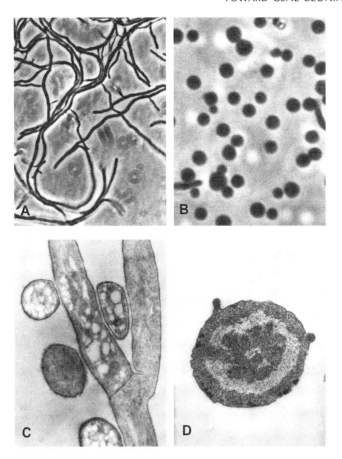

Figure 4.3. *Streptomyces* protoplasts. (A) Light microscope image of mycelium. (B) Light microscope image of protoplasts. (C) Electron micrograph of sectioned mycelium. (D) Electron micrograph of a sectioned protoplast. (A and B, photographs by David Hopwood; C, courtesy of Hans-Ruedi Wildermuth; D, courtesy of Ian Stevenson.)

determined by genes on the same chromosome. The established technique to bring about cell fusion used infection by a virus that caused the membranes to coalesce, but this was not suitable for all combinations of cells and was in any case rather cumbersome. Through his hobbies of alpine gardening and flower photography, Ponte kept an eye on the botanical literature and had spotted reports on the use of PEG to fuse plant protoplasts. He and Anne Hales at the Imperial Cancer Research Foundation had applied the technique to animal cells and had come up with a set of optimal conditions for the treatment, which Ponte told us about before it was published.[5]

Helen prepared protoplasts from two *S. coelicolor* strains both lacking the native sex plasmids, SCP1 and SCP2, and each carrying two or three different auxotrophic mutations. The lack of plasmids was important, because this meant that natural mating did not occur, so any recombinants we obtained must have resulted from protoplast fusion. Helen fused the protoplasts, choosing all the variables exactly as Ponte had recommended: the concentration of the PEG, its molecular weight, other chemicals in the fusion solution, and the time of treatment. She than spread the fused protoplasts on a regeneration medium that lacked the dietary supplements required by the parent strains. In parallel, cultures were made on a medium on which the parental types could grow. When we looked at the plates a couple of days later, we thought there had been a mistake with the composition of the media: there were almost as many colonies on the medium that would select recombinants as on those where the parents could grow. But repeat experiments confirmed that everything was correct. Protoplast fusion was amazingly efficient: almost half of all the individuals emerging from the experiment were recombinants, a far higher frequency than in the *Bacillus* experiments, as Ponte had predicted.[6] Later we varied all possible conditions, but we never improved on the recombination frequency in that first experiment.

Why did protoplast fusion give such striking results? Okanishi and his colleagues had shown that regenerating protoplasts produced a large sac, from which new thread-like hyphae grew and everything returned to normal (Figure 4.4). We postulated that the chromosomes in the sac would replicate to fill it, giving many copies of each of the parental chromosomes and therefore many opportunities for crossing-over, not just a fleeting chance for this to happen as in a natural mating. Our analysis of the recombinants arising from individual fusions told us that all combinations of the parental genes appeared among the progeny, so evidently protoplast fusion brought together complete chromosomes from the two parents, an unprecedented situation in bacterial genetics.

We wondered whether the high recombination frequencies achieved by protoplast fusion would make it easy to create new combinations of genes affecting antibiotic production, as a tool for breeding superior strains. Dick Baltz at the Eli Lilly Company in Indianapolis, who invented protoplast fusion in industrial streptomycetes independently of our work, was motivated by a desire to do just that. Consider the diagram in Figure 4.5. In a strain improvement program, the highest producer among

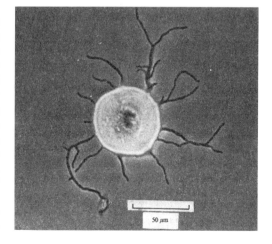

Figure 4.4. A regenerating *Streptomyces* protoplast. The protoplast developed into a large sphere, about 20 times the diameter of the original protoplast, and normal-looking hyphae are now growing out from it. (From Okanishi, M., Suzuki, K. and Umezawa, H. [1974]. Formation and reversion of streptomycete protoplasts: cultural condition and morphological study. *Journal of General Microbiology* 80, 389–400, reproduced by permission of the Society for General Microbiology.)

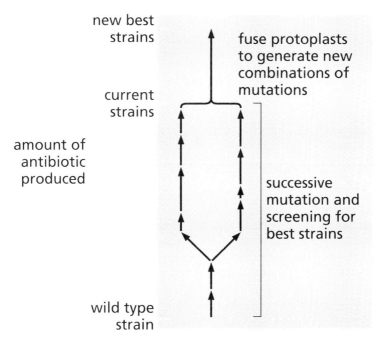

Figure 4.5. A strain-improvement pedigree in which two high-yielding strains were obtained by many steps of mutation and screening, then combined by protoplast fusion to generate higher-yielding recombinant strains.

the survivors of a mutagenic treatment is picked and used in turn as the starting point for the next round of mutagenesis. There is usually a single line of strains leading from the starting culture to the current best strain. In the figure, there are two such lines, leading to two best strains with almost the same level of antibiotic production. However, because the production level depends on hundreds of genes, the mutations in the two branches of the pedigree are most unlikely all to be in the same genes. If it were possible to scramble the genes of the two best strains, a huge number of new combinations of yield-enhancing mutations would result, and most likely some of them would be superior to either of the current best strains. It seemed that protoplast fusion would be the ideal way to do the gene scrambling. We had even shown that protoplasts of four differently-marked strains of *S. coelicolor* could fuse to yield recombinants inheriting genes not just from two parents but from three or even all four, so several lines of selection could perhaps be scrambled together to combine their beneficial genes rapidly into a "super-strain."

Protoplast fusion was certainly used for strain improvement in pharmaceutical companies in the 1980s, but no details were published, which was quite frustrating. That is the downside of working on topics of potential utility. Without inside knowledge, there is often no way of knowing that one's work is being applied commercially. Recently, however, a report brought joy to the hearts of *Streptomyces* geneticists disappointed by the dearth of quotable examples of the practical use of their craft. In

2001, scientists at Maxygen, a Californian company that made its name by devising methods for scrambling DNA outside the cell and then reintroducing it by transformation—a technique they call gene shuffling—picked up on our experiment and took it to the next obvious stage.

The Maxygen scientists pooled the progeny from an initial fusion of our four strains and fused them together again for several further rounds. At the end of the experiment, the proportion of progeny inheriting marker genes from all four of the original parents had gone up dramatically. Next, in collaboration with the Eli Lilly Company, they turned their attention to *Streptomyces fradiae*, which makes tylosin, an antibiotic that Lilly had sold in large quantities for animal use, having developed high-yielding strains by classic strain improvement methods from a low-yielding strain called SF1. Maxygen tested 22,000 survivors of a mutagenic treatment of SF1 for productivity and chose the best 11, which they subjected to a mass protoplast fusion. One thousand progeny were screened, and the best seven were fused together again. Out of 1000 of the resulting progeny, two were as good as or even better than the Lilly production strain. In 1 year, by screening only 24,000 strains, the same end result was achieved as in 20 years of classic strain improvement involving the testing of about a million cultures.[7] Such is the potential of recombination compared with mutation alone as a means of driving evolution. It is no wonder that almost all forms of life have sex, or at least alternatives to it. Ponte was right in predicting the power of gene exchange to improve industrial strains (page 76).

We were pleased with protoplast fusion as a new tool for *Streptomyces* genetics, but it lacked the power of recombinant DNA technology for analyzing and manipulating genes by cloning them. We needed to learn how to make recombinant DNA in *Streptomyces*. We were greatly encouraged in this endeavor by Stan Cohen (Figure 4.6), one of the pioneers of the recombinant DNA revolution, who was with us on sabbatical from Stanford in 1976. We did not achieve our objective until later, but his input was crucial for the success of the project. I regarded it as an accolade that such a high-profile scientist should have chosen to spend time in my laboratory, and we all learned a great deal by exchanging ideas with him.

The Art of Cloning Genes

A typical recombinant DNA experiment, as pioneered at Stanford, starts by breaking open the cells of an organism and purifying the DNA by standard biochemical procedures. The DNA (Figure 4.7) is then manipulated in some way, typically by cutting the molecules with enzymes that recognize a precisely defined sequence, often six base pairs long and symmetrical about its center. The most useful class of enzymes cut the two strands at offset positions, producing ends that are "sticky" because they have single-stranded projections carrying unpaired bases ready to form hydrogen bonds with a complementary strand. Sticky ends made by the same enzyme all have the same sequence, so they can be joined in new combinations (Figure 4.8A). The recombinant DNA then needs to be introduced into a new bacterial host. Foreign genes, with no corresponding DNA sequences in the host, cannot be inserted into its chromosome by normal recombination, so they are spliced into a plasmid,

Figure 4.6. Stanley Cohen in his laboratory at Stanford University, 1997 or 1998. (Courtesy of Stanley L. Cohen.)

which acts as a vehicle or vector for them, replicating the foreign DNA as part of the genetic endowment of the organism (Figure 4.8B).

E. coli, in spite of being the workhorse of bacterial genetics, does not have the advantage of natural genetic transformation, but Stan Cohen and his colleagues, building on an earlier result by Morton Mandel and A. Higa using bacteriophage DNA, induced the bacterium to take up plasmid DNA from the medium in the presence of a high concentration of calcium and after a series of temperature shocks.[8] The mechanism is still not clear, and the best frequency was only 1 in about 100,000 bacteria, but it was adequate when combined with a powerful selection for the transformed individuals.

We could easily have used this procedure to clone *Streptomyces* genes on a vector into *E. coli,* but this was not the aim; we wanted to harness the power of recombinant DNA to manipulate *Streptomyces* genes in their natural host. When the method used for *E.* coli failed with *S. coelicolor,* we decided to try treating protoplasts with PEG in the presence of DNA, reasoning that some DNA molecules might be trapped between the fusing protoplasts and end up inside the regenerating cells.

Because many naturally occurring plasmids of *E. coli* and its relatives carry antibiotic resistance genes, they could be used to select the rare bacteria that had taken

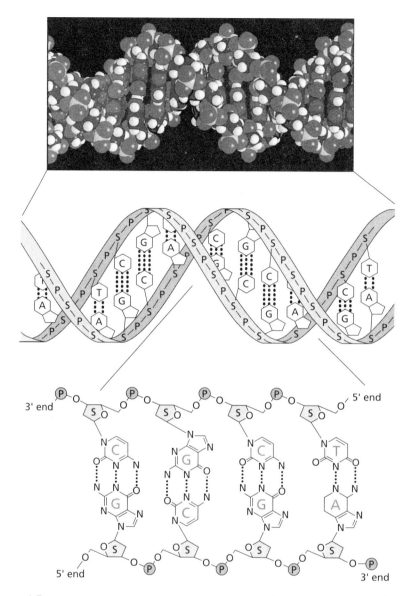

Figure 4.7. The structure of DNA, with a space-filling model at the top, a schematic of the sugar-phosphate backbones and base pairs in the middle, and more chemical detail below. (Space-filling model reproduced by permission from Alberts, B. (2002). *Molecular biology of the cell*, 4th ed. London: Taylor and Francis.)

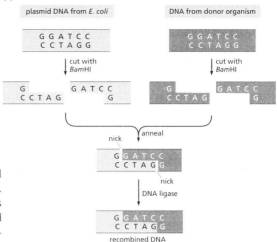

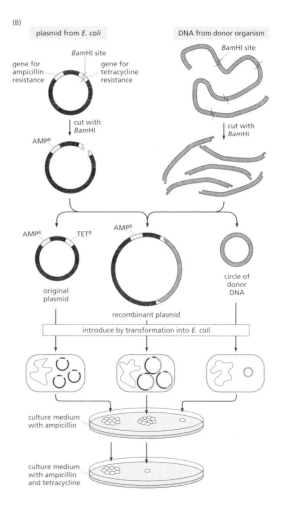

Figure 4.8. Steps in a typical genetic engineering experiment. (A) A restriction enzyme such as *Bam*HI cuts plasmid DNA and donor DNA at the same recognition sequence to generate "sticky" ends that can come together in a new combination and then be joined with the enzyme DNA ligase to make a recombinant DNA molecule. (B) A recombinant plasmid is made by inserting a donor DNA fragment into the *Bam*HI site in the tetracycline-resistance gene of the plasmid, thereby inactivating it; when introduced into *Escherichia coli* by transformation, the recombinants give rise to colonies on ampicillin-containing medium and are distinguished from colonies containing the unchanged plasmid by being unable to grow on tetracycline (bacteria acquiring circles of donor DNA also fail to give colonies). (*Note on nomenclature of restriction enzymes*: the first letter of the genus name is followed by the first two letters of the specific epithet, in italics, then by any strain details, and finally a Roman numeral to indicate whether this is the first or a later enzyme to come from that strain; for example, *Bam*HI comes from *Bacillus amyloliquefaciens*, strain H, and is the first enzyme to be found in this strain.)

up DNA in the Stanford experiments. Hardly any naturally occurring *Streptomyces* plasmids carry antibiotic resistance genes, but Mervyn Bibb had found a powerful alternative way of detecting plasmid-carrying cells. He saw a narrow zone of inhibition, which he called a "tramline," where a strain lacking a plasmid grew in contact with a culture containing SCP2 (Figure 4.9A). When individual plasmid-carrying spores were added to a confluent mass of plasmid-free cells, the equivalent of a tramline formed around each plasmid-carrying colony. Such "pock" formation probably reflects a temporary inhibition of recipient growth when a plasmid is newly transferred into it and copies of the plasmid are spreading within the recipient mycelium. It is characteristic of *Streptomyces* plasmids. The two examples shown in Figure 4.9B and C were found by Tobias Kieser, who arrived in Norwich in 1979 with a PhD from the Swiss Federal Institute of Technology in Zürich and stayed until he left the JII to become a high school teacher in 2004.

In an attempt to demonstrate transformation, Mervyn Bibb added SCP2 DNA to a preparation of fusing *S. coelicolor* protoplasts. The experiment was a spectacular success. Pocks appeared on the regeneration plates, and he calculated that there was a pock for every hundred thousand or so plasmid molecules, as high a frequency as the best reported for artificial transformation of *E. coli*.[9]

We soon discarded the original idea that the protoplasts were just taking up DNA as they fused with each other, because the optimal concentration of PEG for transformation was so low that it caused very little fusion of protoplasts. Instead, it seems likely that PEG causes blebs to form on the surface of the protoplast, and DNA molecules are taken up when they collapse back into the membrane. (A couple of small membrane blebs may even be seen on the surface of an untreated protoplast in the electron micrograph in Figure 4.3D.) Whatever the mechanism, it is a surprising process, because it allows DNA molecules many cell diameters in circumference to cross a membrane that normally excludes even small molecules.

Although many *Streptomyces* plasmids replicate independently of the chromosome, some have evolved a perfect strategy for stable inheritance by being part of the chromosome. *S. coelicolor* has one called SLP1, which occasionally "loops out" and multiplies autonomously. Like SCP1 and SCP2, SLP1 is a sex factor that promotes mating and its own transfer to a new host, such as the closely related *Streptomyces lividans*, whereupon it either reinserts itself into the chromosome or replicates autonomously at a copy number of several per chromosome. This kind of integrating plasmid is confined to the actinomycetes, where several examples have been found since Mervyn discovered SLP1. Their integration into and excision from the chromosome occurs by crossing-over between short sequences on plasmid and chromosome, similar to integration and excision of the F plasmid in *E. coli* (Chapter 3, Figure 3.13).

The interacting sequences on the F plasmid and host chromosome are identical, but in other situations they are different and are recognized by a specialized site-specific recombination enzyme that can break and rejoin DNA molecules very efficiently within such "attachment sites." The classic example is when a bacterial virus integrates into the host chromosome to generate a prophage. SLP1-like plasmids are especially well adapted, because the sites they use for integration are part of the DNA sequences for the transfer RNAs that carry amino acids to the messenger RNA while it is being trans-

TOWARD GENE CLONING 95

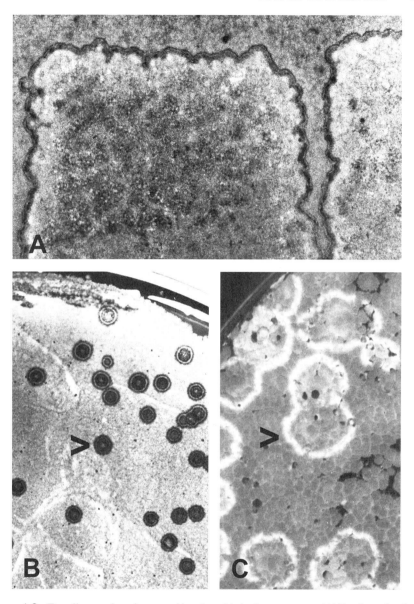

Figure 4.9. Tramlines and pocks caused by plasmids in *Streptomyces*. (A) Patches of a plasmid-containing culture growing on a background of a plasmid-free strain are surrounded by narrow parallel lines ("tramlines") where the background growth has been retarded. (Courtesy of Mervyn Bibb, John Innes Centre.) (B) and (C) Individual plasmid-containing spores have produced "pocks" on a background of a plasmid-free strain; the two panels show the different size and appearance of pocks caused by two different plasmids, and each arrowhead points to an individual pock. (Courtesy of Tobias Kieser, John Innes Centre.)

lated into protein on the ribosome. These are almost identical in all bacteria, so SLP1 can integrate into the chromosomes of a wide range of hosts.

The first *Streptomyces* cloning experiments with plasmids, published in 1980, used the freely replicating SCP2 and the integrating SLP1.[10] Mervyn Bibb was by then a postdoctoral fellow with Stan Cohen at Stanford (he later returned to the JII as an independent member of the *Streptomyces* group), and Charles Thompson held a similar position in my laboratory, assisted by Judy Ward. Charles has gone on to make important discoveries in Geneva, Paris, Basel, and now Vancouver. We shall hear more of both of them later in the book. They used antibiotic resistance genes from *Streptomyces* species as selectable markers, To do so they introduced random fragments of chromosomal DNA from the donor species into *S. lividans* on SLP1 and selected for resistance to the corresponding antibiotic. In this way, they obtained clones carrying the resistance genes that had served to protect the producers from killing by their own antibiotics, and those that Charles isolated proved crucial for the future development of *Streptomyces* cloning vectors.

Soon, the worldwide *Streptomyces* research community was developing plasmid vectors of many kinds and using them to analyze lots of interesting genes. When the object was simply to make a foreign protein for biochemical experiments, a high-copy-number plasmid was best, because, usually, the more copies of a gene, the more of its protein product is made. Tobias Kieser found one he called pIJ101 in another close relative of *S. coelicolor* at a copy number of several hundred per chromosome, and many other such high-copy-number plasmids were found in other laboratories. If instead the objective was to study the physiological effects of the introduced genes, a low-copy-number plasmid such as SCP2 was preferred, because multiple copies of an extra gene often lead to abnormal effects. Integrating plasmids, like SLP1 and another called pSAM2 discovered by Jean-Luc Pernodet at the University of Paris-Sud, are even more suitable, because foreign genes carried into the host chromosome have a one-to-one ratio with host genes. Cloning vectors based on temperate bacteriophages have the same advantage, and they were soon developed.

Using Viruses to Clone *Streptomyces* Genes

Streptomyces phages can easily be isolated using a technique adapted from the study of *E. coli* phages. A spoonful of soil is shaken in water, then passed through a coarse filter, such as a wad of cheesecloth, to remove soil debris, followed by a bacteriological filter to eliminate soil microbes. A little of the liquid is added to molten agar containing a dense suspension of *Streptomyces* spores and spread over a base plate. During overnight incubation, the spores germinate and phages infect the first hyphae they meet, liberating progeny phages that repeat the process for several rounds. When the culture is examined, usually the next morning, a small "plaque," devoid of bacteria, is seen in the creamy "lawn" of bacterial growth. Virulent phages yield clear plaques, because every infected bacterium is killed; temperate phages produce turbid plaques, in which most bacteria are killed but a few survive and grow with the prophage in their genome, conferring immunity to infection by the same virus (Figure 4.10). Phage particles can be picked up in their thousands by touching a sterile

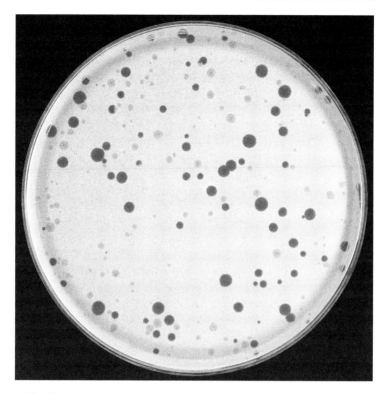

Figure 4.10. Plaques arising from infection of a *Streptomyces* culture by bacteriophages of various kinds, giving large or small, clear or turbid plaques. (Courtesy of Celia Bruton, John Innes Centre.)

needle to the plaque. If you were to do this simple experiment with soil from your own garden you would almost certainly discover phages, and they would probably be new to science, such is the abundance of hitherto undescribed actinomycetes and their viruses.

Figure 4.11 shows an electron micrograph of a *Streptomyces* phage called ΦC31. It was discovered by Natalia Dimitriovna (Natasha) Lomovskaya at the All-Union Research Institute of Genetics and Selection of Microorganisms (VNII Genetika) in Moscow. This institute had developed from a laboratory set up under the leadership of Sos Isaakovich Alikhanian (1906–1985) in 1958 within the Kurchatov Institute of Atomic Energy, and moved to its own premises in 1968. Alikhanian founded the Soviet school of genetics and selection of microorganisms, although he had started out teaching philosophy at Moscow State University and switched to genetics only in 1932, initially working with fruit flies and poultry.[11] Russian genetics had been strong in the 1930s, especially evolutionary genetics, but all this changed after Trofim Denisovich Lysenko's Lamarckian ideas—the inheritance of acquired characters— won out over what was branded "Western mendelian genetics" (in which genes were remarkably stable, only occasionally mutating at random to give new gene combi-

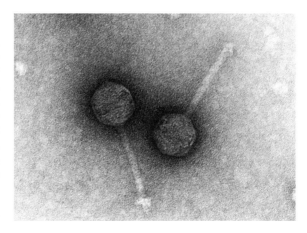

Figure 4.11. Electron micrograph showing two particles of the *Streptomyces* bacteriophage ΦC31 with polygonal head, tail, and base plate that attaches to the cell wall of a host bacterium. The dark material around the phage heads is a metal salt to enhance visibility of the viruses. (Courtesy of Tim Clayton, John Innes Centre.)

nations, which would be favored by natural selection if beneficial). Lysenko's claims, in part supported by faked experiments with plants, found favor with Stalin and the Soviet Communist Party. Not only did he promise rapid progress in breeding the new plant varieties that Soviet agriculture needed, but if acquired human traits, such as altruistic behavior, could be passed on to future generations, the social engineering demanded by Marxism-Leninism could be greatly accelerated.

When the science of genetics as we know it was officially banned in the Soviet Union in 1948, Alikhanian joined an institute dedicated to developing penicillin production. Its charge was broadened in 1952 to include other antibiotics, and it was renamed the All-Union Scientific Research Institute of Antibiotics, where, in parallel with what was happening in commercial companies in the West during the Golden Age, actinomycetes making useful antibiotics were isolated and developed. The Soviet authorities did not accept Waksman's name of *Streptomyces* for these microbes, perhaps because he was seen as having betrayed Russia by emigrating to the United States in 1910 (or perhaps because he was Jewish, as were many of the Russian proponents of mendelian genetics before World War II), so for a long time they referred to them as *Actinomyces*, adding to confusion about their classification. Natasha Lomovskaya was Alikhanian's student, and, like many Russian scientists of her generation, she acknowledges a great debt to his leadership and encouragement.

Alikhanian started looking for genetic exchange in *Streptomyces* in Moscow at the same time as the Sermontis in Rome and I in Cambridge. However, whereas we used strains of *S. coelicolor* as academic models, Alikhanian chose *Streptomyces rimosus* and *Streptomyces aureofaciens,* which make the important tetracycline antibiotics. Like the Sermontis, he hoped to bring the power of genetics to bear on strain improvement. Alikhanian was invited to participate in a symposium held under the auspices of the New York Academy of Sciences in New York on January 6, 1959, at which all the groups who had initiated genetic studies with *Streptomyces* were due to speak for the first time—the others being Giuseppe Sermonti from Rome; Gaylen Bradley, who had started in Lederberg's laboratory but by then had moved to the

University of Minnesota; Hiuga Saito from Tokyo; Waclaw Szybalski from Waksman's institute at Rutgers (who organized the meeting); and me. Unfortunately, Alikhanian was not allowed to attend, but his manuscript was included in the published proceedings.

It was another 9 years before I met Alikhanian (Figure 4.12), at an international symposium on Genetics and Breeding of *Streptomyces* held in Dubrovnik in the former Yugoslavia at the end of May 1968. The great advantage of holding such meetings in Eastern Europe in those days was that scientists from the Soviet bloc could usually participate. My wife and I took our three small children to the meeting, and we spent time with Alikhanian, by then a benevolent grandfather figure, who enjoyed carrying our 7-month-old daughter along the beach. He told me that he had just set up a laboratory for *Streptomyces* bacteriophage genetics and had asked Lomovskaya to lead it. At the Dubrovnik meeting, a group of scientists from the Czechoslovak Academy of Sciences Institute of Microbiology in Prague, who were pioneering work on antibiotic biosynthesis, developed the idea of a series of international symposia on the Genetics of Industrial Microorganisms (GIM), to be held every 4 years. They organized the first meeting in Prague in August 1970, a courageous act, given that this was the second anniversary of the Soviet invasion and genetics was still regarded by some in the Eastern bloc as a misguided Western subject.

Natasha Lomovskaya was expected at the Prague meeting, where she was due to give a paper on phage ΦC31. Unfortunately, she had been ill with pneumonia and could not get the necessary government documentation of good health in time to come, so another member of the Russian delegation, Alexander Boronin, read

Figure 4.12. S. I. Alikhanian in 1966. The dedication reads: "To dear Natasha on my sixtieth Birthday 26/XI/66 S. Alikhanian." (Courtesy of Natalia Lomovskaya.)

her paper. Alikhanian handed me a letter from Natasha in which she introduced ΦC31 and asked for some of my genetically marked derivatives of *S. coelicolor* to study the phage further. Returning to Norwich, I was happy to supply the strains she had requested, and I asked her for ΦC31 and a host strain she called "*Actinomyces coelicolor* 66" that the Moscow group used in their work. Later, the Russian strain was renamed *Streptomyces lividans* 66, though it is in fact a very close relative of *S. coelicolor*.

As was the rule in the Soviet Union, they had published a first paper on ΦC31 in Russian[12] and only the following year in a Western journal in English, the international language of science. They described how the phage produced plaques on strain 66. At first, they thought it originated from a prophage in the chromosome of *S. coelicolor*, just as lambda was found when it excised from the *E. coli* chromosome in Lederberg's laboratory and produced plaques on another strain. Later, it turned out that ΦC31 had infected *S. lividans* by chance from a different, unidentified source.

I finally met Natasha (Figure 4.13) at the 13th International Congress of Genetics in Berkeley in 1973, where she gave her paper only 2 hours after arriving. The Russian delegation had to fly from Moscow to Rome, then on unconnected flights to Brussels, New York, across the United States with two changes, finally arriving in San Francisco after a 48-hour journey. She was so tired after reading her paper that she could not keep her eyes open during my presentation, so we had dinner together on Telegraph Avenue to exchange results and ideas. She was a bit nervous being among hippies on her first trip to the United States, but I assured her they were en-

Figure 4.13. Natasha Lomovskaya in conversation with Keith Chater at the Sixth International Symposium on the Biology of Actinomycetes in Debrecen, Hungary, August 26–30, 1985. The author's wife, Joyce, is on the right. (Courtesy of the University Medical School, Debrecen.)

tirely friendly. She reminded me recently that I had to write everything down because she still had trouble understanding spoken English.

My meeting with Natasha was the beginning of an ongoing friendship during which my wife and I came to know her and her husband Leonid Fonstein, who worked on *E. coli* phages in Moscow, better and better. She is a slight figure, deeply thoughtful but with a twinkle in her eye and a dry sense of humor that seems to have found increasing expression since she and Leonid emigrated to the United States after *perestroika* to work at the University of Wisconsin, Madison, until their retirement to California in 2003. We reflected recently on the huge change in her life, after constant difficulties in Moscow, of eventually becoming a U.S. citizen, a step that was unimaginable during the Soviet period.

In spite of restrictions, Natasha was able to come to Sheffield in 1974 for the second in the series of GIM meetings, but after that she could not travel to the West until 1980; no reason was ever given. In January of that year, the policy mysteriously changed, and she and her former student, Valeri Danilenko, were allowed to take up my invitation to spend a couple of months in Norwich, where she worked with my colleague Keith Chater. Keith had joined me in Norwich in 1969, after doing a PhD thesis in Birmingham on transduction in *Salmonella*, and liked working with phages. His group had turned to a study of ΦC31 shortly before Natasha's visit, and he had been in correspondence with her. He went to Moscow for 2 weeks in September 1979 to write a first review of the genetics of *Streptomyces* phages, which they completed during Natasha's visit to Norwich. After that, Keith concentrated on using ΦC31 for gene cloning. He has always admired Natasha's painstaking and insightful studies on the *in vivo* genetics of the phage, which provided an indispensable platform for his genetic engineering work. Keith and a Spanish postdoctoral fellow, Juan Suarez, used ΦC31 for one of the first cloning experiments in *Streptomyces* in parallel with those of Mervyn Bibb and Charles Thompson using plasmids.[13]

Later Keith's group, especially his long-time assistant Celia Bruton, developed ΦC31 into a whole series of specialized cloning vectors that became powerful tools for *Streptomyces* gene isolation and analysis. In building them, they were guided by the example of the lambda phage of *E. coli*, in which nonessential genes had been eliminated to make room for foreign genes and many other subtle genetic tricks had been applied. The ability of ΦC31 vectors to infect the *Streptomyces* mycelium, instead of requiring protoplasts to introduce DNA into the host, is a notable advantage. An even simpler trick for introducing DNA into *Streptomyces* was developed later and soon became a routine tool in *Streptomyces* genetics.

This very straightforward technique was based on the amazing discovery that *E. coli* carrying the F plasmid will mate with almost any kind of cell, including those of yeasts and plants. *Streptomyces* comes within its range, as Philippe Mazodier and colleagues at the Pasteur Institute in Paris reported in 1989.[14] Complicated genetic constructs that may require several successive steps of cutting and splicing DNA are often made in *E. coli*, using so-called bifunctional vectors consisting of part of a plasmid from *E. coli* and another part from a *Streptomyces* vector, which enables them to replicate in both hosts. In this way, everything goes much faster, because *E. coli* cultures develop more rapidly than those of *Streptomyces* and do not need to go through a sporulation stage to produce a new set of cells for each cloning operation.

The next step, that of extracting the DNA from *E. coli* and transforming *Streptomyces* protoplasts with it, can be neatly sidestepped by mixing. *E. coli* carrying a sex plasmid with *Streptomyces* mycelium on a plate of growth medium, whereupon DNA is transferred. After a day or so, two antibiotics are added: one to kill the *E. coli* and a second, to which the vector confers resistance, to leave only the *Streptomyces* cells that have received the plasmid to develop into colonies.

All the technical developments described in this chapter meant that, by the end of the 1980s, *Streptomyces* could be genetically manipulated using a wide range of natural and artificial processes. New combinations of genes could be made by natural mating or by protoplast fusion, as well as by the much more powerful technique of cloning genes and reintroducing them into the host via protoplast transformation using plasmids, phage infection, or mating from *E. coli*. As a result, genes controlling key *Streptomyces* traits were identified and isolated as segments of DNA in many laboratories. They included sets of genes for synthesizing the basic building blocks of the cells, utilizing complex food sources, making antibiotics, and controlling steps in the development of the colonies. In parallel, the architecture of the chromosome was gradually revealed, as described in the next chapter.

5

From Chromosome Map to DNA Sequence

By the mid-1980s, many questions still remained about the *Streptomyces coelicolor* chromosome. Was it linear or circular, and how big was it? Many genes had been cloned piecemeal, but how were they arranged on the chromosome, and how did this reflect their functioning? What about all the genes that had not been studied? Fortunately, entirely new methods for analyzing chromosome structure were invented around this time. In 1987, Georges Carle and Maynard Olson at Washington University, St. Louis, and Cassandra Smith and Charles Cantor at Columbia University, New York, discovered how to physically map DNA molecules as large as whole bacterial genomes and eukaryotic chromosomes; it came to be called pulsed field gel electrophoresis (PFGE).[1] Meanwhile, genetics was undergoing a momentous change with another development, whole genome sequencing, pioneered by Craig Venter at The Institute for Genomic Research (TIGR) in Rockville, Maryland.

This chapter describes how these new technologies revolutionized our understanding of *Streptomyces* genetics. Having a complete inventory of the organism's genes began to reveal how *Streptomyces* bacteria are adapted to life in the soil, including making antibiotics. These are topics for later chapters. Here I describe how the *S. coelicolor* chromosome went from a virtual map with fewer than 150 genes to a DNA sequence of more than 8 million base pairs, representing almost 8000 genes, and how even a cursory look at the content of the genome already told us a lot about how the organism coped with its environment.

Mapping the *Streptomyces* DNA

Helen Kieser (Figure 5.1) began to apply the PFGE technique to the *S. coelicolor* chromosome soon after the method was announced. Adapting the published protocols, she embedded samples of mycelium in little blocks of agarose gel and broke open the cells by immersing the blocks in a lysozyme solution to degrade the walls, followed by a proteolytic enzyme to digest much of the cellular contents. The fragile chromosomal DNA floated out into the agarose block, protected by the gel from even slight shearing forces that would break the DNA if it were liberated from the mycelium into a liquid. The next step was to cut the DNA into pieces with a restriction enzyme that recognizes a six-base-pair sequence consisting only of A and T bases. Such sequences are rare in *Streptomyces,* because its DNA contains an unusually low

Figure 5.1. Helen Kieser talking about physical mapping of the *Streptomyces* chromosome during a practical course at the Huazhong Agricultural University, Wuhan, China, April 1998. (Courtesy of Tobias Kieser, John Innes Centre.)

proportion of A and T compared with G and C (G and C make up about 72% of the bases, compared with about 50% in *Escherichia coli* and 41% in humans), so a manageable number of fragments are produced that can be separated from each other according to size.

The standard method for separating small DNA molecules is to put a DNA solution in a slot at one side of a square agarose gel and immerse the gel in a salt solution with electrodes at either end of the tank. When the current is switched on, the DNA fragments move toward the positively charged electrode (DNA, being an acid, is negatively charged) at a speed inversely related to their size. This separates molecules up to about 20,000 base pairs long, such as those made by cutting a plasmid or phage DNA with a restriction enzyme that recognizes a frequently occurring site. But larger molecules do not separate, because, soon after the current is switched on, they straighten to lie in the direction of the electric field and then slide like eels through the channels in the gel at a rate independent of their length.

The PFGE technique (Figure 5.2) uses two pairs of electrodes. When one pair is switched on, the DNA molecules start to move toward the positive charge, the smallest molecules fastest. Even very large molecules start to separate from one another, but they soon align themselves with the electric field and no further separation occurs. The current is now switched to the second pair of electrodes, at an angle to the first, and the molecules reorientate themselves, the smallest fastest, so they separate further. By repeating the current switching for many hours or even a few days, DNA molecules up to millions of base pairs long are separated. The fragments are visualized by staining the DNA with a dye that fluoresces under ultraviolet light and taking a photograph.

When Helen started the work, only two enzymes were available to cut at sites containing only A and T bases: *Dra*I from *Deinococcus radiophilus* cuts at 5´TTTAAA3´, and *Ssp*I from a *Sphaerotilus* species recognizes 5´AATATT3´. But neither seemed suitable for *S. coelicolor* DNA. *Ssp*I cut the chromosome into too many fragments to analyze, and *Dra*I did not seem to cut the DNA reproducibly. Just then, in September 1987, Keith Chater and I, with our wives, made a trip to Leningrad and Moscow. Natasha Lomovskaya was our host, and of course we visited her institute. One of the scientists there told us about a new restriction enzyme they had discovered, called *Vsp*I after a *Vibrio* species, with a recognition site that was another permutation of three As and three Ts: 5´ATTAAT3´. Could we have the bacterium that produced it, or at least a sample of the enzyme for Helen to try out? He thought it doubtful, but would check with a higher authority.

Keith and I were already late for our next appointment at another institute, but we hung on as long as we could in the lobby of the building under the watchful eye of the party commissar. Just as we were about to give up, our newfound friend came bounding down the stairs and shook Keith heartily by the hand as we took our leave. In the car, Keith examined a little plastic tube of liquid lodged between his fingers: it was labeled "VspI." The enzyme survived for several days in Keith's luggage without being refrigerated and, back in Norwich, Helen used it with great success, but pretty soon the small amount of enzyme was exhausted. Keith wrote to the head of the Moscow institute, asking for the producing strain or a sample of the enzyme

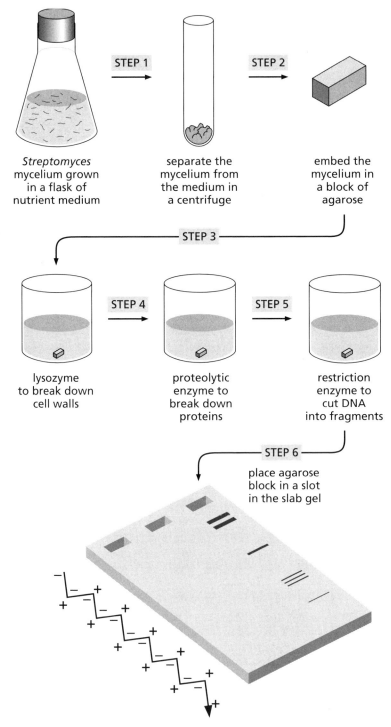

Figure 5.2. Electrophoretic separation of DNA molecules by pulsed field gel electrophoresis (PFGE).

(he didn't admit we had tried their enzyme, just indicating why we thought it would be useful). He was given the name of an institute in Siberia that owned the strain. It didn't look promising but, just in the nick of time, New England Biolabs of Beverly, Massachusetts, the largest supplier of restriction enzymes, discovered another enzyme, which they named *Ase*I after *Aquaspirillum serpens*, that seemed to have the same DNA target site as *Vsp*I. A friend at the company sent us a free sample, and it worked beautifully. Soon we learned from French colleagues working on another *Streptomyces* species at the University of Nancy how to vary the conditions to get *Dra*I to work, so Helen could now move forward with the physical mapping of the *S. coelicolor* chromosome using the two restriction enzymes.

Helen found that *Dra*I generated just 8 chromosomal fragments, and *Ase*I produced 17. Next, she determined the order of these fragments in the chromosome, to create the beginnings of a physical map with the restriction sites on it. She used a combination of two approaches. In the first, she cut the DNA with each enzyme separately and ran the fragments on two different PGFE gels. She lifted the bands representing individual DNA fragments produced by one of the enzymes (for example, A1 through A7 in Figure 5.3) from the gel, separated the two strands (i.e., "denatured" the DNA) by heating, and converted each DNA sample into a "probe" by tagging it with a radioactive isotope of phosphorus. She denatured the DNA *in situ* in the second gel and transferred the pattern of single-stranded fragments (B1 through B7 in Figure 5.3) by "blotting" onto a square of plastic pressed against the gel. (This is a technique called Southern blotting after its inventor, Edwin Southern of the University of Edinburgh; it became so pervasive in molecular biology that when variations of the technique were invented to study RNA and protein, they were called Northern and Western blotting.) On soaking the plastic sheet in a solution of the first probe (e.g., A1), any DNA on the plastic with the corresponding sequence formed a double strand ("hybridized") with it. Excess probe solution was washed away, and the hybridizing bands could be recognized when the plastic sheet was exposed to X-ray film. This served to identify the overlaps between the A and B fragments.

The second approach depended on a trick that Tobias Kieser invented to isolate small pieces of the chromosome, called linking clones (AL1 through AL7 and BL1 through BL7 in Figure 5.3), each carrying an *Ase*I or a *Dra*I site (it is a bit complicated to go into but is described in the reference[2]). When each linking clone was used as a probe against the set of PFGE fragments generated with the same enzyme, it hybridized only with the two fragments on either side of that particular restriction site, thereby establishing their order. The resulting map seemed to show that the chromosome was a circular molecule, because Helen found linking clones that connected adjacent fragments all around the chromosome without a gap, finally resolving the ambiguity between a linear and a circular chromosome that had existed for the previous 25 years.

Helen could then relate her physical map to the existing genetic map of the *S. coelicolor* chromosome by making probes corresponding to genes that had been both genetically mapped and isolated as clones, and hybridizing them to gels containing the sets of fragments generated with each of the two restriction enzymes. Each gene hybridized with just one *Ase*I fragment and one *Dra*I fragment, confirming the overlap between them and locating the gene on the segment common to the two fragments. We were very pleased when the order of the genes on the genetic map, and

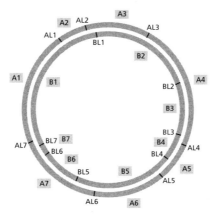

Figure 5.3. Mapping a bacterial chromosome by PFGE. In this hypothetical example, two enzymes happen to cut the chromosome each in seven places. The outer ring shows the sites where enzyme A cuts, labeled AL1 through AL7, producing fragments A1 through A7, and the inner ring shows the corresponding outcome for enzyme B. Each of the A fragments is isolated and used as a probe to hybridize against a gel carrying the set of B fragments to reveal overlaps; for example, fragment A7 would hybridize to B5, B6, B7, and B1, showing that they are partially or entirely contained within A7 and must form an adjacent set, though in an unknown order. These B fragments are now hybridized to a gel carrying the A fragments, whereupon B6 and B7 would hybridize only to A7; therefore, B6 and B7 must be entirely contained within A7 and adjacent to each other. B1 would hybridize with A1 as well as A7, proving that A1 flanks A7 on one side; and B5 would hybridize with A6 as well as A7, proving that A6 flanks A7 on the other side. (Small clones carrying AL1 through AL7 and BL1 through BL7 are the linking clones referred to in the text, which allow the order of the A and the B fragments to be determined.)

even their approximate distances apart, were confirmed by this physical analysis. The whole chromosome was a DNA molecule of about 8 megabases (Mb; 1 Mb = 1 million base pairs), one of the largest known in any bacterium. Evidently the fact that *Streptomyces* seems more complicated than simple rod-shaped bacteria like *E. coli* reflects a larger endowment of genes, as described in detail in subsequent chapters.

Just after Helen's article on the physical mapping of the circular *S. coelicolor* chromosome appeared in the spring of 1992,[2] a good friend of ours, Carton Chen from the National Yang Ming Medical University in Taiwan (Figure 5.4), arrived to spend a sabbatical summer working with Helen. On the first day in the laboratory, he exploded a bombshell. He told us about experiments in his group suggesting that Lomovskaya's *Streptomyces lividans* had a linear chromosome: they could detect a free chromosome end. Chromosome linearity was almost unprecedented for a bacterium, so Carton and Helen did more experiments to try to resolve the matter, and by the end of the summer they had isolated both chromosomal ends and proved their structure. It was clear that Carton had been absolutely right: *S. lividans* had a linear chromosome.[3] Soon, Helen established that this was true for *S. coelicolor* as well.

Where had the earlier analysis gone wrong? It turned out that Helen's physical mapping had been compromised by a totally unexpected feature: the same sequence

Figure 5.4. Carton Chen at the John Innes Centre, Norwich, September 1998. (Courtesy of Tobias Kieser, John Innes Centre.)

of DNA, about 20,000 base pairs long, occurs at the two ends of the chromosome, in opposite orientation, forming "terminal inverted repeats," or TIRs. The two restriction fragments representing the ends of the chromosome carried the TIRs, and so they both hybridized to a particular linking clone and appeared to be joined, but it was a false linkage.

The next step in analyzing the *S. coelicolor* chromosome was taken by Matthias Redenbach (Figure 5.5), a German postdoctoral fellow who came to Norwich from the University of Kaiserslautern in 1993, bringing with him a collection of pieces of the *S. coelicolor* chromosome cloned in a special kind of vector called a cosmid. Cosmids carry the unique sequence of phage lambda DNA that is the signal for packaging it into a virus particle during natural infection in *E. coli*. This allows the DNA to be artificially packaged in the test tube using a mix of viral proteins to give virus particles that are equivalent to transducing phages. They can inject their DNA into *E. coli*, where it replicates because the vector contains part of an *E. coli* plasmid. A special advantage of cosmid vectors is that the cloned DNA is long and almost constant in size, because it has to fit exactly inside the phage head.

With help from colleagues in Kaiserslautern, Norwich, and the laboratory of Haruyasu Kinashi in Hiroshima, where Matthias went next, he picked out from the large pool of cosmids an ordered set that included the whole chromosome. He did this by choosing a cosmid at random and using its chromosomal insert as a probe against the pool of cosmids, whereupon it hybridized with clones that overlapped with it on one side or the other. With these as new probes, Matthias could identify adjoining cosmids, and so on progressively from one end of the chromosome to the other. He picked a set of 319 cosmids that just covered the chromosome, called a tiling path because they overlap like tiles on a roof.

Next, Helen plotted the positions of more than 150 genes on the set of clones. To do this, she obtained DNA samples from laboratories around the world representing

Figure 5.5. Matthias Redenbach at the 12th International Symposium on Biology of Actinomycetes (ISBA'91), Vancouver, August 2001. (Courtesy of Tobias Kieser, John Innes Centre.)

the genes that people had been busy cloning. Such was Helen's popularity among the *Streptomyces* community, stemming from her bubbling enthusiasm, outgoing personality, and interest in everyone's professional (and private) lives, that almost everyone responded. She used each gene as a probe against the set of clones, causing it to hybridize either to a single clone or to an overlap between two adjacent clones, as expected. The result was a detailed genetic and physical map that the *Streptomyces* community started to use routinely in their research (Figure 5.6). They would e-mail Helen asking for clones or groups of adjacent clones carrying genes they were interested in, and she filled dozens of requests. Even more important, the set of ordered DNA fragments provided the material to sequence the entire genome, a milestone in *Streptomyces* genetics and in my own career. The project was carried out at the Sanger Centre on the Wellcome Trust Genome Campus at Hinxton, 6 miles south of Cambridge.

Sequencing the Streptomyces Chromosome

One sunny July morning in 1997, Helen, Tobias, and I drove the 60 miles from Norwich to Hinxton. Safely wedged in the back of the car was a Styrofoam box of dry ice containing small plastic tubes of frozen *E. coli* cells carrying some of the cosmid clones in Matthias's minimal set. After several years of committee meetings and grant applications, the money had been found and the long-awaited *S. coelicolor* genome-sequencing project was about to begin. As we picked up our visitor's badges from the security desk at the Hinxton campus (Figure 5.7), behind us was a stained glass window with a representation of a tree of life, as well as a string of the letters A, T, C, and G: a fragment of genetic code (Color Plate 3).

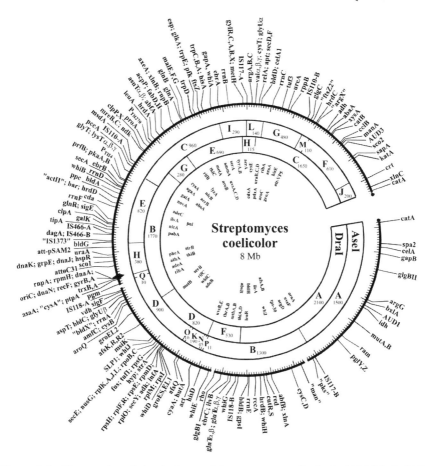

Figure 5.6. The genetic and physical map of *Streptomyces coelicolor* in 1996. The bold line represents the chromosome, with *oriC* represented by the diamond shape in the middle and the terminal proteins shown as dots at the ends. On the outside are genes mapped to the set of cosmid clones covering the entire chromosome in a "tiling path," except for three gaps (shown as interruptions in the bold line). Inside the chromosome are the positions and sizes (in kilobases) of fragments generated by two restriction enzymes, *Ase*I and *Dra*I. In the center are the approximate positions of genes mapped earlier from the results of crosses.

Watson and Crick published their model for the structure of DNA in April 1953, but already their minds were racing ahead toward the goal of cracking the genetic code implicit in the model.[4] The order of bases along the DNA molecule would carry the information for assembling in specific order the string of amino acids that makes up each protein, and it seemed that just 20 amino acids needed to be coded for. With only four kinds of bases, evidently the code could not be a one-to-one relationship, nor two-to-one, with two adjacent bases corresponding to each amino acid, because that would give only 16 combinations. The minimum number of bases for each amino acid would be three, a triplet. This would offer 64 combinations, or codons, which was obviously

Figure 5.7. The Sanger Centre (renamed Sanger Institute in 2001) on the Wellcome Trust Genome Campus, 9 miles south of Cambridge, at dusk. The campus covers 55 acres (22 hectares) of park and estate land surrounding a mid-18th century mansion, Hinxton Hall. (Courtesy of Richard Summers, Sanger Institute.)

more than enough to code for 20 amino acids. Soon, genetic and biochemical studies not only proved a triplet code but also cracked it. All 64 codons are used, 61 for amino acids and three for stop signs. Two amino acids have a single codon, one has three, many have two or four, and three have six. Table 5.1 gives the details of the code, which is the same, except for one minor change, in all forms of life.

A burning issue in thinking about the code had been the question of punctuation. Were there "commas" separating the triplets of bases that formed the codons, or were the codons defined simply by reading the bases off in threes starting at one end of the gene? Francis Crick, Sydney Brenner, and colleagues at the Laboratory of Molecular Biology in Cambridge, which had been founded after Watson and Crick's pioneering work in the Cavendish Laboratory (the physics department of the university), performed a brilliant genetic experiment with an *E. coli* bacteriophage called T4, which gave an unambiguous answer: the code is comma-less.[5] The experiment showed, not surprisingly, that deletion or addition of a base pair in the DNA of a particular phage gene abolished its function, giving rise to a special appearance of the plaques it made on a lawn of bacteria. Two deletions or additions also inactivated the gene but, crucially, a combination of three deletions or additions fairly close together in the gene restored the normal function. So did just one deletion (or addition) closely followed by an addition (or deletion). The conclusion was that the translation machinery, as it reads the code in the messenger RNA from one end, goes into the wrong "reading frame" when it encounters a missing

Table 5.1. The genetic code

First position (5'- end)	Second position				Third position (3'- end)
	T	C	A	G	
T	TTT Phe (F)	TCT Ser (S)	TAT Tyr (Y)	TGT Cys (C)	T
	TTC Phe (F)	TCC Ser (S)	TAC Tyr (Y)	TGC Cys (C)	C
	TTA Leu (L)	TCA Ser (S)	TAA Stop	TGA Stop	A
	TTG Leu (L)	TCG Ser (S)	TAG Stop	TGG Trp (W)	G
C	CTT Leu (L)	CCT Pro (P)	CAT His (H)	CGT Arg (R)	T
	CTC Leu (L)	CCC Pro (P)	CAC His (H)	CGC Arg (R)	C
	CTA Leu (L)	CCA Pro (P)	CAA Gln (Q)	CGA Arg (R)	A
	CTG Leu (L)	CCG Pro (P)	CAG Gln (Q)	CGG Arg (R)	G
A	ATT Ile (I)	ACT Thr (T)	AAT Asn (N)	AGT Ser (S)	T
	ATC Ile (I)	ACC Thr (T)	AAC Asn (N)	AGG Ser (S)	C
	ATA Ile (I)	ACA Thr (T)	AAA Lys (K)	AGA Arg (R)	A
	ATG Met (M)	ACG Thr (T)	AAG Lys (K)	AGG Arg (R)	G
G	GTT Val (V)	GCT Ala (A)	GAT Asp (D)	GGT Gly (G)	T
	GTC Val (V)	GCC Ala (A)	GAC Asp (D)	GGC Gly (G)	C
	GTA Val (V)	GCA Ala (A)	GAA Glu (E)	GGA Gly (G)	A
	GTG Val (V)	GCG Ala (A)	GAG Glu (E)	GGG Gly (G)	G

Abbreviations for amino acids: Ala, alanine; Arg, arginine; Asn, asparagine; Asp, aspartic acid; Cys, cysteine; Gln, glutamine; Glu, glutamic acid; Gly, glycine; His, histidine; Ile, isoleucine; Leu, leucine; Lys, lysine; Met, methionine; Phe, phenylalanine; Pro, proline; Ser, serine; Thr, threonine; Trp, tryptophan; Tyr, tyrosine; Val, valine.

or extra base, but returns to the correct reading frame when it passes a compensating change, either a single mutation of the opposite kind (e.g., an addition compensating a deletion), or two further mutations of the same kind, making three altogether. It was the latter observation that revealed that the reading frame represents a triplet of bases.

The string of letters in the stained glass window was

TGATAATAGTTCAGGGAAGATTGATCCGCAAACGGTGAGCGTTAATGATAG

Reading this from the left and translating it gave the following sequence:

STOP, STOP, STOP, phenylalanine, arginine, glutamic acid, aspartic acid, STOP, serine, alanine, asparagine, glycine, glutamic acid, arginine, STOP, STOP, STOP

In single-letter code for the amino acids (Table 5.1). this became:

... FRED.SANGER ...

Fred Sanger is one of just two scientists to win two Nobel Prizes in science, the other being Marie Curie (Linus Pauling won one prize for chemistry and another for peace). Sanger's first prize was in 1958 for inventing the means to sequence proteins; the second, in 1980, was for sequencing DNA, arguably the biochemical technique that has left the greatest legacy in biology.

As we left the lobby of the Hinxton facility and entered the Sanger Centre, above the reception desk was an illuminated display with an endless sequence of A, T, G, and C streaming across it. This was the pooled real-time output from dozens of automated sequencing machines working away, 24 hours a day, sequencing the DNA of whatever organisms were in play at the time—human, worm, malaria parasite, *Mycobacterium tuberculosis*, or whatever. Soon, *S. coelicolor* DNA would start appearing there. We were met by Bart Barrell, head of the Pathogen Sequencing Unit at the Sanger Centre (the Wellcome Trust had assigned the *S. coelicolor* project to this unit because, although not a pathogen, it kills pathogens), and Mike Quail, who was in charge of the DNA that was used to carry out the sequencing projects. He took the precious clones Helen had brought and checked them into their freezer space in the "clone hotel."

David Harris ran the team that determined the *S. coelicolor* sequence. His group worked on the clones in a series of stages. First, the DNA was purified from the *E. coli* cultures we had brought. It was then fragmented with bursts of ultrasound into pieces 1500 to 2000 base pairs long, and these were cloned again in *E. coli*. Plasmid preparations were made from about 1000 random colonies. This served to bulk up the DNA to make enough for sequencing.

The method that Fred Sanger invented to sequence DNA depends on copying single-stranded "template" DNA, using the DNA polymerase enzyme that replicates a bacterial chromosome. Four reactions are run in parallel, each containing the four bases of DNA attached to the sugar deoxyglucose, with a triphosphate group added, making a building block called a nucleotide. In each reaction, a small fraction of one of the nucleotides is replaced by a chemically modified form of the natural compound, called a dideoxy derivative. Now and again, the polymerase incorporates this analogue of the normal nucleotide into the growing chain, because the 5´ sugar carbon carries an –OH group (Figure 4.7) and therefore is available for joining to the preceding building unit. But this stops further chain extension, because the 3´ carbon lacks an –OH group and therefore cannot receive the next unit needed to extend the chain, as explained in Figure 5.8A and B. This means that in, say, the "C" reaction, some of the daughter molecules stop short at all the various positions where C occurs in the sequence, and likewise for the other three reactions.

In an early version of the technique, the DNA molecules generated in these reactions were separated by electrophoresis in different tracks in a long slab made from a material called polyacrylamide. The molecules were made radioactive so that they could be seen by exposing an X-ray film to the gel slab (Figure 5.8D), in the same way as in the PFGE technique described earlier. Nowadays, the DNA fragments are made to fluoresce instead of being radioactive, using a dye attached to the blocking nucleotide, with a different color for each of the four reactions. A mixture of all four reaction products is run down the gel on an automated sequencing machine; the machine can detect the length of each fragment by its time of travel through the gel, and whether it represents an A, T, G, or C reaction by its fluorescent color. The machine compiles all the lengths as a sequence for up to 500 bases, or even more under the best conditions, especially in current machines, where the DNA samples travel down the gel in individual narrow tubes, rather than being next to each other on the same slab. The machine can be loaded up to three times in a 24-hour period.

FROM CHROMOSOME MAP TO DNA SEQUENCE 115

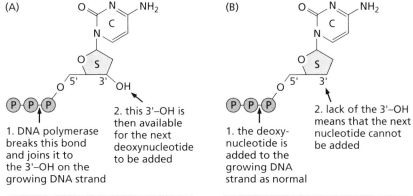

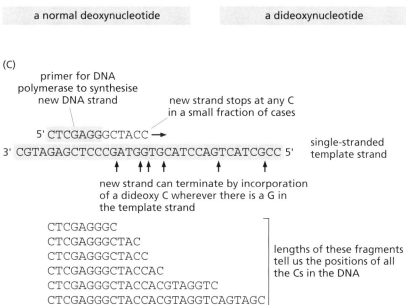

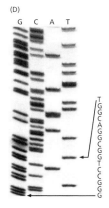

Figure 5.8. Fred Sanger's method of DNA sequencing. (A) and (B) A normal and a dideoxy version of a cytosine nucleotide. (C) Incorporation of a dideoxycytosine in various positions during copying of the template strand causes the strand to stop growing at that position *(arrows)*. (D) A segment of a sequencing gel with a "read-out" of part of the sequence. (Courtesy of Paul Hoskisson, John Innes Centre.)

The output from the machines was fed to a computer that assembled the sequences of the 1000 *S. coelicolor* clones from each cosmid by matching overlaps between clones and finding complementary matches between opposite strands—both strands being represented in the sequence with equal probability. After assembly, there were usually just a few gaps in the sequence, which were closed by obtaining further sequences over the critical regions. The sequence was declared finished when it was judged to be "greater than 99.99% accurate," based on prior experience of sequencing millions of base pairs of DNA in other projects. Looking at it the other way round, there was less than 1 error in 10,000 base pairs, or 10 typical genes. Such is the skill of the Sanger teams that the sequence was routinely much better than that.

When foreign DNA replicates many times in a host such as *E. coli*, it can undergo deletions and rearrangements so that the order of bases differs from the true sequence in the original host. This is always a worry when sequencing projects use cloned DNA, as ours did; today, the standard technique is a so-called shotgun approach, in which DNA is isolated directly from the original source and fragmented without amplification in *E. coli*. Therefore, in a first strategy meeting to discuss the *S. coelicolor* project, John Sulston, head of the Sanger Centre and soon to win a Nobel Prize for his pioneering work on genome analysis, suggested we start with 10 adjacent clones, to check their quality. This had the added advantage that the pilot study would give a realistic estimate of the cost of the entire genome sequence before the BBSRC signed a large contract. Over the next few weeks, we held our collective breaths as the first clones were finished. Everything looked great. The expected overlaps between adjacent cosmids in the tiling path were present and correct, and the average size of the chromosomal fragment in each cosmid was exactly as expected, so it was unlikely that the clones had lost DNA. This meant that, by early 1998, the full project could go ahead.

The clones averaged 37,500 base pairs long and overlapped by about 12,500 base pairs, so each clone contributed 25,000 base pairs of unique sequence. In practice, the overlaps were quickly recognized and finished sequence was determined on only one of the overlapping clones. Otherwise, one third of the entire sequencing effort would have been redundant. There originally appeared to be just three gaps in the tiling path of cosmids along the chromosome, but there turned out to be more—this almost always happens in such projects because of chance resemblances between sequences, giving false overlaps—so extra clones had to be isolated to bridge the gaps. And the original estimate of the genome size had been about 8 Mb, but it ended up at almost 8.7 Mb, because the unusually high proportions of G and C caused the DNA to run a bit faster than expected in the PFGE technique, resulting in underestimation of the lengths of the fragments.

Recognizing the Genes

Analyzing a stretch of newly sequenced DNA, called annotation, proceeds in two stages. The sequence is first scanned with computer programs that recognize the characteristic features of genes. A major concern is to identify the start-point of each gene. Any gene that encodes a protein must begin with a codon for incorporating a chemical variant of the amino acid methionine, called *N*-formylmethionine, which is used to initiate protein synthesis. This start codon is usually ATG, but ATG also codes for methionine at internal positions in genes, so it does not automatically define the start-point. Moreover, other start codons can code for *N*-formylmethionine, notably GTG, which in internal positions in genes is one of the valine codons. With its high GC content, more genes start with GTG in *Streptomyces* than in most organisms, making the start-points of the genes harder to spot because the annotators have always to consider both ATG and GTG as potential start codons. Help is often available in the form of a special sequence, a few base pairs before the start-site, which is characteristic of bacterial genes. This ribosome binding site allows the messenger RNA to dock with the ribosome, helping it to recognize the start codon and begin translating the message into protein from the correct point.

After each start codon is a string of a few hundred codons for the amino acid sequence of the protein—any combination of the 61 in the code—followed by one of the three kinds of stop codons: TAA, TAG, or TGA. The pattern of codons between start and stop can indicate that a real gene has been correctly recognized. This is because of the degeneracy of the code—the fact that almost all the amino acids have more than one possible codon, with most of the degeneracy in the third codon position (Table 5.1). *Streptomyces*, with high GC DNA, chooses predominantly codons with G or C in the third codon position, and there are characteristic frequencies of G and C in the first and second positions too. *Streptomyces* genes are therefore a bit easier to identify than those from most other genomes; in an organism with 50% G and C, the frequencies of different synonymous codons are more random. Because there are no commas in the code, a stretch of DNA could be read starting at any of three positions, and in either direction (one strand is the coding strand at some places on the chromosome, and the other strand plays this role in other regions), so there are six possible reading frames for any segment of DNA. Only one of the six fits the expected GC pattern in the different codon positions. Using these criteria, the computer can predict the positions of genes with a high degree of confidence (Color Plate 4).

Next comes the interesting bit, when the sequence starts to tell us about the biology of the organism. A lot can be learned by comparing the sequences of newly discovered genes with those already known in other organisms (some striking examples are described in Chapter 6). It is a miracle of modern computer science that the sequence of a new gene can be sent over the Internet to the U.S. National Center for Biotechnology Information and compared with all the tens of billions of base pairs of sequence already deposited in publicly funded databases, using a set of programs called BLAST. In at most a few minutes, a report comes back with the hits listed in order of similarity. Comparing DNA sequences gives information about how genes with a common ancestor may have diverged from each other over evolutionary time. But to predict the functions of new genes, it is usually much more informative to compare the sequences of the proteins the genes would encode. This is because, stemming from the degeneracy of the code, two DNA sequences can diverge a lot while still encoding proteins with very similar or even identical amino acid sequences.

In comparing proteins, the computer algorithms make allowances for evolutionary changes, such as substitution of one amino acid by another with similar chemical and physical properties, a so-called conservative change. Another thing that can happen during evolution is deletion or addition of a few amino acids, so the program can allow for a certain number and size of gaps when two sequences are compared. These various allowances affect the likelihood that two proteins carry out the same job, so there are usually two statistical scores in the report: one for "identity," the degree to which two amino acid sequences match exactly, and a second, higher value for "similarity," allowing for conservative changes and gaps.

Clearly gene finding, and especially the prediction of gene function, has a subjective element, and this is reflected in the annotations that are submitted to the sequence databases. Words like "putative," "probable," and "possible" abound, and the quantitative likelihood scores are recorded for people to interpret as they see fit. Any annotation is a work in progress and will be constantly confirmed, extended, or overthrown, either by further computer comparisons as the databases expand or, better, by new experimental work. After all, it is experiments that provide information on the functions of genes in the first place. Without them, computer comparisons can become increasingly misleading as genes are assigned related functions when they differ progressively from a gene studied experimentally.

As Julian Parkhill began to annotate the *S. coelicolor* genome sequence, and Stephen Bentley and Ana Cerdeño completed the task, about 50% of the genes they recognized resembled genes already described in the genome of some other organism and assigned a function there, but to varying degrees. At one extreme, the function of a new gene can be identified unambiguously from such comparisons. For example, the sequences of the more than 50 proteins that make up the bacterial ribosome are highly conserved. They have to be, because ribosomes have become adapted over eons of evolution to function like a well-oiled machine, and any except conservative changes in the protein sequences are likely to upset its smooth working. Comparison of a *Streptomyces* gene product to one of these, from almost any bacterium, could give a similarity value greater than 60%, essentially proving the function of the new gene. In other situations, two proteins may have only a modest similarity end-to-end but share a conserved string of amino acids that represents the active site of an enzyme. From this motif we might predict that our new gene encoded, say, a phosphatase that would remove a phosphate group from some molecule, but not what that molecule was or what its biological role might be. Or the new gene might encode a protein that had all the hallmarks of a membrane protein (see Chapter 7), but we could not necessarily predict its role there.

Another 30% or so of the predicted *S. coelicolor* genes resembled some previously sequenced gene but with no assigned function, so this did not help us predict its role; such genes are called "conserved hypothetical." Finally, the remaining 20% showed no convincing degree of resemblance to any previously sequenced gene: they were just "hypothetical." With 50% "known," 30% "conserved hypothetical," and 20% "hypothetical," the outcome was typical of genome projects in other organisms.

The Sanger Centre (renamed Sanger Institute in 2001), reflecting the policy of the Wellcome Trust that funds it, insists that all genome sequences it generates are publicly accessible. This has huge advantages for the progress of science because, not only does it avoid wasteful overlap of effort, but it allows everyone to take advantage of new knowledge as soon as it is obtained. In stark contrast, a publicly available genome sequence for the dangerous pathogen *Staphylococcus aureus* was delayed for years; word got around that several commercial companies were doing it, but none of the sequences was published.

When the *S. coelicolor* project started, a web site was established on which data began to accumulate. All DNA sequences more than 1000 base pairs long were displayed there, with nightly updates. The progress of the project was recorded, as each cosmid proceeded from random sequencing of fragments, through finishing, to annotation. When annotation of a cosmid was completed, the results were immediately deposited in the public databases with a link from the project web site, and an e-mail club was established, with a couple of hundred subscribers who were sent hot-off-the-computer updates of the latest annotations and could submit their own comments. Sometimes people were lucky enough to find relevant gene sequences early in the project; others had to wait longer for their favorite genes to appear in the database as a later cosmid was sequenced. Piecemeal sequencing, which had gone on in dozens of laboratories for years, often occupying much of the total project time of a PhD student or a postdoctoral scientist, gradually stopped and people could begin asking interesting biological questions almost from the start of their work, using the sequence in the database.

The *S. coelicolor* sequencing project, which cost about 2 million dollars, was funded by the BBSRC at a rate that would allow it to be completed in 3 years, but in the end it took almost 4. Apart from the need to fill gaps in the cosmid tiling path and to sequence a slightly larger genome than expected, the Sanger Institute became caught up in the race to complete the human genome project as an open-access resource, in competition with commercial sequencing in the United States, so effort had to be diverted away from the sequencing of microbial genomes. Nevertheless, the last base pair slid into place on July 21, 2001, to a great sigh of relief all round. The sequence turned out to be 8,667,507 base pairs long, and the Sanger annotators had recognized 7825 protein-encoding genes in it, the largest number found in any completed microbial genome sequence at that time. Included in the contract was the sequence of the linear SCP1 plasmid, 256,073 base pairs encoding 353 proteins. The SCP2 plasmid, a circle of 31,317 base pairs, was completed later by putting together stretches of sequence obtained by several laboratories in different countries.

A bacterial genome sequence is rather different from that of a higher eukaryote, including the human genome. For these huge genomes, a so-called draft sequence, with most of the genes annotated but falling short of having every base pair in place, is often published first; only later is the genome completed, including long stretches of repetitive DNA that is both very laborious to sequence and of limited interest compared with the coding DNA. Bacteria do not have this "baggage." Nor are their genes interrupted by long stretches of noncoding DNA, the so-called introns, which are spliced out from the RNA transcripts of higher eukaryotes to give the uninterrupted coding sequences of the messenger RNA molecules. Bacteria have so little noncoding DNA that, on average, genes occupy just over 1000 base pairs all along the genome. In the human genome, the figure is more like 100,000 base pairs per gene, but a gene can sometimes encode several different proteins because of alternative splicing out of introns.

A complete genome sequence is like an Aladdin's Cave: it is so full of goodies that one becomes dazzled, as described in Richard Burton's classic 19th-century translation of *The Arabian Nights*[6]:

> Aladdin . . . returned to the garden where . . . [the trees bore] for fruitage costly gems; moreover each had its own kind of growth and jewels of its peculiar sort; and these were of every colour, green and white; yellow, red and other such brilliant hues and the radiance flashing from these gems paled the rays of the sun in forenoon sheen . . . Aladdin walked amongst the trees and gazed upon them and other things which surprised the sight and bewildered the wits; and, as he considered them, he saw that in lieu of common fruits the produce was of mighty fine jewels and precious stones, such as emeralds and diamonds; rubies, spinels and balasses, pearls and similar gems astounding the mental vision of man.

What are some of the jewels and precious stones that the *S. coelicolor* genome sequence revealed? One new insight relates to an aspect of *Streptomyces* genetics that had occupied several laboratories for years. The chromosomes of many species of *Streptomyces* grown under laboratory conditions or in fermenters were found to change in structure, often by losing long stretches of DNA. After genome linearity was demonstrated, it emerged that these deletions occurred principally at the ends of the chromosomes. Up to a million base pairs of DNA might be lost from either or both ends, but the organism still survived—in the laboratory. Some variants had become sensitive to an antibiotic, or sporulated poorly, but they usually grew at a nearly normal rate and were surprisingly healthy, considering they had lost a quarter of their genes.

This remarkable finding was powerfully illuminated by completion of the *S. coelicolor* genome sequence. The end regions are just as full of normal-looking genes as any other part of the genome, so they are not "junk" DNA. However, the genes in these regions are evidently not essential for life, otherwise strains lacking them could not survive, even in the laboratory. In fact, if we plot the positions of obviously essential genes, such as those encoding a complete set of ribosomal proteins, transfer RNAs, and enzymes that load amino acids onto the transfer RNAs, as well as those that replicate the DNA and transcribe it into RNA, none fall in these end regions. In contrast, genes encoding enzymes for breaking down starch, cellulose, or chitin, for example, are found predominantly within a couple of megabases of either end. We called these regions the "arms" of the chromosome and the central part the "core". The left arm is about 1.5 Mb long, and the right arm 2.3 Mb, leaving just under 5 Mb for the core.

The arms of the chromosome carry many genes that are adaptive under particular conditions in the organism's natural habitat, not all the time. For example, genes encoding cellulases are useless if the organism is not trying to grow on this carbon source, but they contribute to its long-term survival by giving it great flexibility to adapt to life in the soil. During a press conference marking publication of the complete *S. coelicolor* genome sequence,[7] Stephen Bentley of the Sanger Institute likened the genome to a Boy or Girl Scout who took seriously the motto: "Be prepared!" The core is the scout, while the arms are the backpack containing a coil of rope, Swiss army knife, compass, and everything else a well-prepared scout might need from time to time.

Why So Many Genes?

The most striking feature of the *S. coelicolor* sequence is its size and the corresponding number of genes it encodes: almost twice as many as *E. coli*, which has 4289 genes.[8] The *S. coelicolor* sequence even contains more genes than yeast, a eukaryotic fungus with about 6000 genes. This came as a big surprise considering that, in general, eukaryotes are expected to be more complex than prokaryotes. But not all actinomycetes have so many genes. *Mycobacterium tuberculosis,* with 3924 genes,[9] similarly to *E. coli*, has about half the number in *S. coelicolor*.

The Sanger Institute has developed a marvelous piece of free computer software called the Artemis Comparison Tool (ACT), which compares the DNA sequences of two entire genomes and plots the matches on a graph according to their positions on the chromosomes, with a dot for each sequence match. When this is done between *S. coelicolor* and *E. coli*, no recognizable picture emerges. The two genomes diverged so early from their last common ancestor, maybe 2.2 billion years ago, that there is no significant conservation of gene order. However, the comparison between *S. coelicolor* and *M. tuberculosis* is revealing. In this comparison, we see a random-looking pattern of dots, telling us that many similar genes in the two organisms now occupy different relative positions. However, superimposed on this "noise" is a clear diagonal line of dots representing genes that are still arranged in the same order, inherited *en bloc* from an ancestor common to the two actinomycetes. Interestingly, this conserved arrangement, while covering the whole length of the *M. tuberculosis* chromosome, corresponds just to the core region in *S. coelicolor*, not the arms. We conclude that genes in the arms were acquired from elsewhere in a process called horizontal gene transfer. At some stage in the process, the *Streptomyces* chromosome became linear, perhaps by merging with a linear plasmid that gave it the ability to replicate in linear form. This idea is supported by the finding that some linear plasmids, such as one called SLP2 in Lomovskaya's *S. lividans* strain, have the same end sequences as the host chromosome, probably reflecting a recent exchange of ends between them.

Horizontal gene transfer is an important concept in evolution.[10] On a recent time scale, it is a major factor in the origin of antibiotic-resistant pathogenic bacteria (by the transfer of plasmids into them) during the period since antibiotics were introduced into medicine, but it has probably been operating ever since different life forms began to occupy different niches. Plasmids are important agents for transferring DNA between distantly related organisms, and viruses have played a part too, but mere transfer of the DNA is not enough.

Once foreign genes have entered a new host, they face problems if they are to become part of the host chromosome. Homologous recombination leads to exchange of DNA sequences during meiosis in eukaryotes and after chromosome segments have been transferred during mating, transduction, or transformation in bacteria, but, as the name suggests, it requires the participating DNA molecules to be very similar. This is not the case after horizontal gene transfer, so other means are needed to insert the foreign DNA into the chromosome of the new host. They often involve segments of DNA that have gained the ability to jump from one location to another. If these "transposons" carry genes that might benefit a new recipient, they take the useful genes with them when they enter the chromosome, so natural selection acts to retain them. It is significant that the ends of the *S. coelicolor* chromosome arms contain a lot of transposons. Some look functional, but many are incomplete relics of successive insertion events occurring on top of each other. I shall talk more about transposons in Chapter 9 as tools to inactivate the genes they jump into.

If the core regions are descended from a common ancestor and the arms have evolved by acquiring inserted DNA, we might expect the arms of different *Streptomyces* chromosomes to be more different from each other than the cores. This was strikingly confirmed with the publication in May 2003 of the second complete sequence of a *Streptomyces* chromosome, for *Streptomyces avermitilis,* the producer of avermectin (the insecticidal antibiotic described in Chapter 2), by a group led by Haruo Ikeda in Satoshi Ōmura's Kitasato Institute.[11] They found a very similar arrangement of related genes in the core regions of the *S. coelicolor* and *S. avermitilis* chromosomes, but the arms contained many genes found in only one of the two genomes.

This chapter has taken a giant leap from the pregenomic phase of *Streptomyces* genetics that lasted into the 1990s to the genomic period at the start of the new millennium. We have learned that *S. coelicolor* has a most unusual chromosome: it is large and linear, is differentiated into core and arm regions, and it has almost twice as many genes as bacteria such as *E. coli* and *M. tuberculosis*. Why should this be? The answer must surely lie in the very different habitats and life styles of the different organisms. *Streptomyces* is a free-living soil bacterium, whereas the others have evolved as fellow-travelers or parasites of animals.

Compared with the environments encountered by specialized pathogens, the soil is an extraordinarily complex habitat in which microbes are exposed to a huge range of stresses of all kinds, physical, chemical, and biological. The temperature, the abundance of water and oxygen, the salinity, and the pH can all vary over a wide range. So can the availability of nutrients. The organisms absolutely require sources of carbon, nitrogen, iron, sulfur, and phosphorus, as well as a series of other elements in tiny amounts, but many of these may be supplied in different forms, and their availability, like the physical factors, varies from place to place and from time to time in the soil. The same applies to the numbers and kinds of other soil organisms, many of which are competitors: hence the capacity of *Streptomyces* to make antibiotics.

Faced with all these challenges, one strategy would be for a microbe to become highly specialized to succeed under just a few of all the possible conditions it might encounter, and many pathogens have done that. At the other extreme, a microbe might develop an extreme flexibility to allow it to survive, and even flourish, under a whole range of different circumstances. *Streptomyces* has evidently opted for this second strategy. In the next two chapters, I consider how its genetic endowment encodes the features that adapt it superbly for living in the soil.

6

Bacteria That Develop

Colonies of rod-shaped or spherical bacteria, such as *Escherichia coli* or a *Staphylococcus*, consist of piles of similar-looking cells. *Streptomyces* colonies (Figure 6.1) are much more complicated, both morphologically and physiologically. In a slice through a colony (Color Plate 5), young foraging mycelium can be seen at the margins, mature vegetative mycelium in the middle, and aerial hyphae giving rise to spores at the top. It is only in the central region of the colony that an antibiotic, the red pigment, is being produced; the other parts lack pigment. The different regions can be described as different tissues, and this is very unusual for a bacterium.

In this chapter, we follow a *Streptomyces* colony as it develops from a spore, changing its morphology in a succession of stages (Figure 6.2), and consider the challenges the organism faces along the way. The spores must survive and be dispersed in the environment, where they need to germinate to produce the vegetative mycelium in appropriate situations. As this mycelium grows, it extends the cell wall at tips and branches, places cross walls in the hyphae in suitable positions, and populates the mycelial compartments with copies of the chromosome and plasmids by replication and by mating with other mycelia. Then, as nutrients in the colony's surroundings become exhausted, the aerial mycelium must be initiated and metamorphose into chains of spores to conclude the life cycle. All this involves genetically regulated microengineering of amazing complexity, revealed through painstaking laboratory work by many people, using a whole range of experimental techniques. This chapter, even on its own, is a good example of the international collaborations that have been a striking and enormously satisfying feature of the development of knowledge about *Streptomyces*.

Figure 6.1. Scanning electron micrograph of a young *Streptomyces* colony on an agar plate. Note the young vegetative hyphae growing from the edge of the colony and the piled-up aerial mycelium in the center. The scale bar is 100 micrometers, or 0.1 mm. (Courtesy of Kim Findlay, John Innes Centre.)

Spore Survival

Although a new *Streptomyces* colony can develop from a mycelial fragment, just as a plant can regenerate from a piece of stem, the spores have the main responsibility for producing the next generation. They lie dormant in the soil until conditions become suitable for mycelial growth; the published record is 70 years, but this is surely not the maximum survival time. Soils contain far more *Streptomyces* spores than growing hyphae. Plants, especially those that inhabit difficult environments such as deserts, face similar challenges, and they too exist mostly in a dormant state, as seeds. When rain falls on the desert, some of the seeds germinate and produce metabolites that inhibit the germination of other members of the same species. This is clearly adaptive. It would be bad for all the seeds to germinate at the first sprinkle of rain, which might not be sustained: better for some to try their luck and others to wait for a later opportunity. The same seems to happen with *Streptomyces*. An autoinhibitor of spore germination has been described that, like those involved in seed germination, acts only on the strain that makes it.[1]

Dormant *Streptomyces* spores must maximize their chances of finding suitable conditions for the next generation. One important factor is to resist being washed into the depths of the soil, where there is not enough oxygen to grow. So *Streptomyces* spores are hydrophobic, floating on water surfaces and tending not to sink below the water table. They are carried back into the top layers of the soil when it becomes waterlogged. This hydrophobicity is seen when water falls on the surface of *Streptomyces* colonies and the droplets run around without wetting the culture, like water on a duck's back (Figure 6.3).

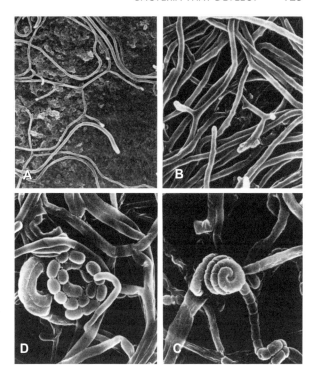

Figure 6.2. Scanning electron micrographs of four stages in colony development in *Streptomyces lividans*. (A) Edge of a colony showing young vegetative mycelium. (B) Mature vegetative mycelium just beginning to produce aerial branches. (C) A young, coiled aerial hypha beginning to sporulate. (D) A chain of mature spores. (Courtesy of Jeremy Burgess, John Innes Centre.)

The basis of the hydrophobicity is quite complex. Electron microscopy shows that the surface of the *Streptomyces* aerial mycelium, and hence of the spores it produces, has a range of structures, with elaborate hairs or spikes in some species (Figure 6.4B,C). In *S. coelicolor,* the spores appear smooth at low magnification (Figure 6.4A), but they are actually covered by a "basket-work" of paired rodlets (Figure 6.5A) encoded by two sets of genes.

A Dutch group led by Lubbert Dijkhuizen at the University of Groningen found a pair of adjacent genes in the *S. coelicolor* genome sequence that encode two hydrophobic proteins, the "rodlins." Two converging studies led to the identification of eight genes encoding another family of hydrophobic proteins, named "chaplins" (from *S. coelicolor h*ydrophobic *a*erial *p*roteins). Marie Elliott, a Canadian postdoctoral fellow in Mark Buttner's group at the John Innes Centre, recognized them among the genes that she found to be transcribed in a wild-type strain but not in a mutant strain that failed to produce aerial mycelium (I describe how this is done in Chapter 9). In another approach, Dennis Claessen in the Groningen group used "reverse genetics"; he purified proteins from the aerial mycelium, determined parts of their amino acid sequences, and asked the computer to find genes in the genome sequence that would encode them. Three of the eight chaplins are probably anchored in the outer layers of the spore wall, and the other five bind to them, making a complex aggregate on the outside of the spores. These proteins belong to a class called amyloids that can self-assemble into filaments (as in the brains of patients with Alzheimer's disease).[2] Formation of the rodlet layer on the spore surface depends on the products of both rodlin and chaplin genes, because

Figure 6.3. Drops of water on the surface of a confluent *Streptomyces* culture. Most of the drops have a "cap" of spores picked up by moving over the culture, but two (second from the left and second from the right) are shiny, with no covering of spores. (Courtesy of Andrew Davis, John Innes Centre.)

it fails to form if either set of genes is mutated.[3] In the chaplin mutant, no fibrils are seen, whereas in the rodlin mutant the chaplins form a less structured layer than in the wild type; this is also seen when purified chaplins are dried down on a plastic surface (Figure 6.5B).

Streptomycetes might perhaps use another stratagem to stay afloat. One of the surprises in the *S. coelicolor* genome sequence was a series of genes resembling

Figure 6.4. Electron micrographs of spores of different *Streptomyces* species. (A) *Streptomyces coelicolor* with smooth spores. (B) *Streptomyces glaucescens* with hairy spores. (C) *Streptomyces viridochromogenes* with spiny spores. (Courtesy of Hansruedi Wildermuth, John Innes Centre.)

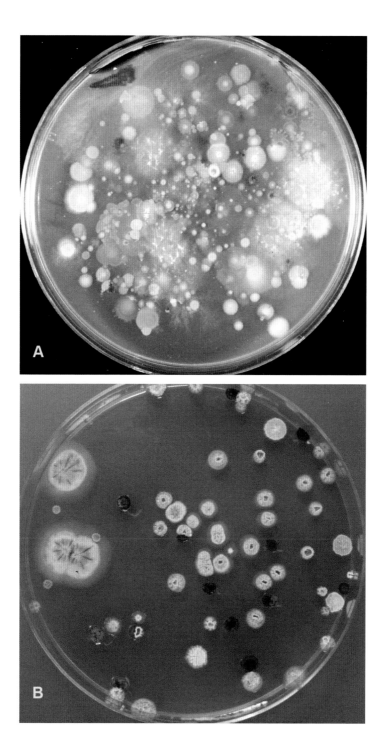

Color Plate 1. (A) A soil isolation plate, with colonies of bacteria, fungi, and *Streptomyces*. (Courtesy of Penny Hirsch, Rothamsted Research.) (B) Colonies of four different *Streptomyces* species. (Courtesy of Helen Kieser, John Innes Centre.)

Color Plate 2. (A) Cultures of different *Streptomyces* species in a 24-well plate. (Courtesy of William Maese, Lederle Laboratories, Pearl River, NY.) (B) *Streptomyces coelicolor* colonies. The blue pigment is the antibiotic actinorhodin that gives the organism its name: *coelicolor* = "sky color." (Courtesy of Andrew Davis, John Innes Centre.)

Color Plate 3. The stained glass window called "The Tree of Life" at the Sanger Institute (window created by Kathy Shaw, photograph by Richard Summers). The real tree trunk seen in the grounds outside, at the bottom of the picture, was felled by lightning in 1999 and later carved by Richard Bray into a sculpture called "Hinxton Spiral," based around a helix, with images of the organisms whose genomes were being sequenced at the Sanger Institute at the time.

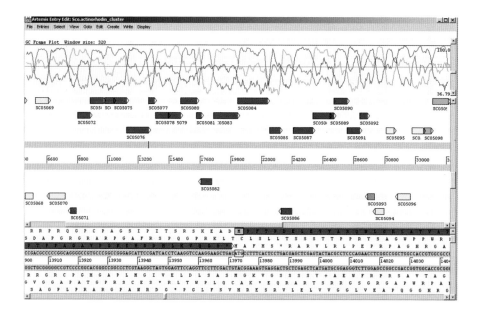

Color Plate 4. Annotation of a stretch of *Streptomyces coelicolor* DNA displayed using the Artemis software. The window is divided into three panels. In the upper one, the three colored lines represent the average G + C content of the DNA in each of the three triplet positions, calculated in a moving segment of 320 bases; notice how the three lines come together at some places, or "nodes," and then separate again as "bubbles"; the bubbles are the genes and the nodes are the gaps between them. In the middle panel, the predicted genes are shown as broad, colored arrows: *red*, previously sequenced genes of known function (in this example, the actinorhodin gene cluster); *yellow*, genes resembling a gene of known function in another genome; *brown*, genes resembling a gene of unknown function in another genome; *green*, genes with no significant resemblance to any other gene. Each gene is in one of the six possible reading frames, three for each strand of the DNA. The lower panel shows details of the DNA sequence at the boundary between two genes, with a translation into amino acids in three-letter code, in all six possible reading frames, the correct one being highlighted in red. The ATG start codon of the downstream gene is boxed in yellow. (Courtesy of Stephen Bentley, Sanger Institute.)

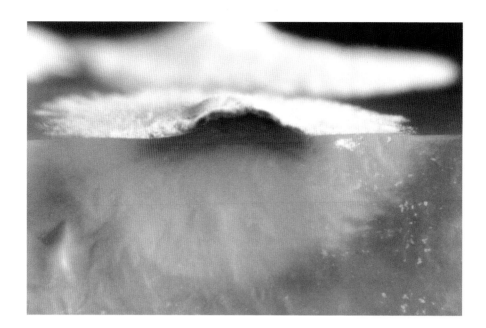

Color Plate 5. Vertical section through a colony of *Streptomyces coelicolor*. Note the vegetative mycelium, mostly below the surface of the agar medium, and the sporulating aerial mycelium above the surface. The red antibiotic is being produced by mature vegetative hyphae in the center of the colony. (Courtesy of Jamie Ryding, John Innes Centre.)

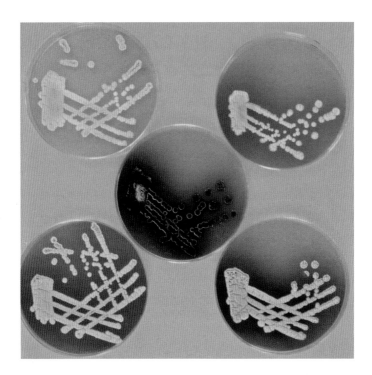

Color Plate 6. A set of *act* mutants of *Streptomyces coelicolor*. The wild type, making blue actinorhodin, is in the middle, surrounded by mutants blocked at different points in the actinorhodin biosynthetic pathway. (Courtesy of Brian Rudd, John Innes Centre.)

Color Plate 7. The squid *Euprymna scolopes* emitting a beam of light generated by *Vibrio fischeri* growing in its light organs inside the transparent mantle. The light helps to make the animal invisible to predators approaching from below, preventing it from being seen against a moonlit sky. (Courtesy of Ned Ruby and Margaret McFall-Ngai; for details, see Ruby, E. G. [1996]. Lessons from a cooperative, bacterial-animal association: the *Vibrio fischeri-Euprymna scolopes* light organ symbiosis. *Annual Review of Microbiology* 50, 591–624)

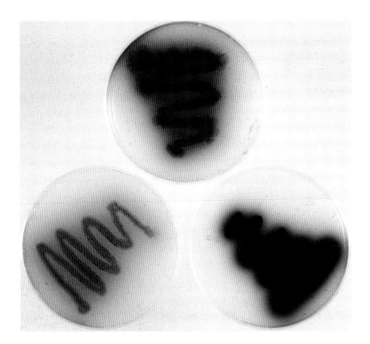

Color Plate 8. A genetically engineered *Streptomyces* culture making the purple, hybrid antibiotic mederrhodin, and parental cultures making blue actinorhodin and brown medermycin. (Courtesy of Helen Kieser, John Innes Centre.)

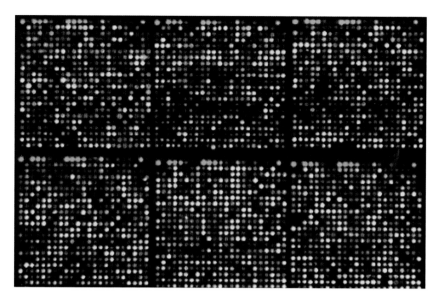

Color Plate 9. Part of a *Streptomyces coelicolor* genomic microarray studying the expression of genes after a heat shock. Genes expressed more strongly at 42°C than at 30°C give red spots, those expressed less strongly at 42°C appear green, and most of the genes, expressed equally at the two temperatures, give yellow spots. The green spots at the top of each block are for calibration. (Courtesy of Giselda Bucca, University of Surrey.)

Color Plate 10. Scab disease of potato caused by *Streptomyces scabies*. Note the characteristic angular lesions caused by the skin tearing back (Tobias Kieser, John Innes Centre, from a specimen provided by Rachel Bircham).

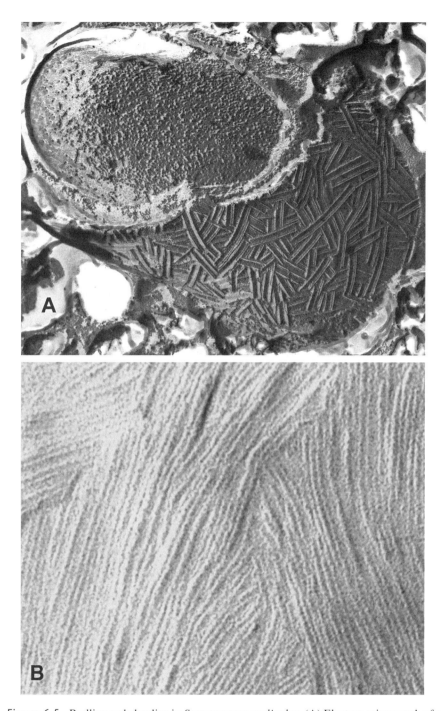

Figure 6.5. Rodlins and chaplins in *Streptomyces coelicolor*. (A) Electron micrograph of a broken spore, showing the rodlet layer on the right in this freeze-fracture preparation and inner layers exposed on the left. (Courtesy of Hansruedi Wildermuth, John Innes Centre.) (B) A layer of filaments formed when a solution of chaplin proteins was allowed to dry on a plastic surface. (Courtesy of Dennis Claessen, University of Groningen.)

those making gas-filled chambers called gas vacuoles that enable various aquatic bacteria to float. Some cyanobacteria use them to hang in the water at a depth where the light intensity is optimal for their photosynthesis. No one has seen such vacuoles in *Streptomyces,* and when Geertje van Keulen in the Groningen group mutated the *S. coelicolor* genes the cells could still float, so perhaps the proteins encoded by these genes have taken on another role.[4] They have special properties, with one side hydrophobic and the other hydrophilic, so they may be dedicated to some process that involves membrane interactions or adhesion. The hunt is on for their real function.

Spore Dispersal

Plant seeds use many devices for dispersal by animals. The seeds may latch onto fur with hooks, or fruits may be attractive to eat so that the seeds pass through the animal's gut and germinate in the dung (as in an *Acacia* species dispersed by elephants). Could *Streptomyces* have evolved something similar for its spores? Keith Chater at the John Innes Centre has speculated that the carriers are earthworms and other soil invertebrates. When the spores are deposited on the soil surface mixed with very fine soil particles as worm casts, they can easily be blown about by the wind. (A sinister analogy is "weaponization" of anthrax spores by mixing them with a fine powder as a carrier in biological warfare.) Recently, Václav Krištůfek at the Institute of Soil Biology in České Budějovice in the Czech Republic found that certain worms related to earthworms are indeed attracted by particular *Streptomyces* species, supporting the idea that worms might selectively feed on the colonies.[5] A group of arthropods called springtails (Collembola) also like to graze on *Streptomyces* cultures, and they too may be important agents of dispersal. Compounds in the outer spore coat (some of them obvious because they are brightly colored: Color Plate 2A) or in the rodlet layers might help to protect the spores from digestion by their invertebrate carriers.

One of the attractants for worms and springtails seems to be a compound called geosmin, which many *Streptomyces* species make. It causes the evocative smell of freshly dug earth. Springtails respond to it, and this would help to ensure that they eat and disperse the *Streptomyces* spores. More arrestingly, camels are said to be particularly good at smelling geosmin, so we could picture a camel in the desert following the scent of geosmin until it reached an oasis, where *Streptomyces* would be growing in the moist soil. As the camel slaked its thirst, *Streptomyces* spores would stick to its muzzle and be carried to the next oasis it visited. This is such a nice idea; let's hope it turns out to be true.

A very special system for dispersing *Streptomyces* spores concerns a group of leaf-cutting ants that cultivate molds as food in "gardens" made from pieces of plant leaves.[6] The ants and the molds have evolved an amazing mutual dependence. Unfortunately for both partners, a parasitic fungus attacks the farmed molds, threatening the survival of the ant population. Enter a *Streptomyces* species that produces an antibiotic which has no effect on a wide range of fungi but which specifically inhibits the parasite. The streptomycete sticks to the outer surface of the ants in specialized regions on female ants—queens and workers—but not on males. This supports the idea that the ants adaptively disperse the streptomycete, because it is the queen that lays the eggs for the next generation of ants and establishes a new fungus garden for them, and it is the workers that tend the garden. It may be that the hydrophobic layers on the surface of *Streptomyces* spores play a role in this by helping the spores stick to hydrophobic surfaces on the ants, but there must be a more subtle interaction to explain the specificity for female ants of particular species.

Spore Germination

How does a spore monitor the environment and germinate only when conditions are favorable? Some *Streptomyces* spores need raised levels of carbon dioxide to germinate. Perhaps this acts as a signal that other organisms have already found conditions favorable for growth and respiration. Carbon dioxide would be a very general germination trigger, but some streptomycetes probably respond to specific signals. These species interact with plants, and their spores may respond to chemical germination triggers emanating from them.

In the layer around roots, called the rhizosphere, many microbes live on nutrients secreted by the plant and on sloughed-off plant cells. At least two examples of commercial products depend on the ability of *Streptomyces* strains to colonize the rhizosphere, where they help control fungal pathogens of greenhouse crops, especially those that can wipe out whole trays of seedlings, or fungi that kill turf grass. One is a strain of *Streptomyces griseoviridis* that the Finnish company Verdera markets as Mycostop,[7] and the other, called Actinovate, contains *Streptomyces lydicus* and is sold by Natural Industries, Inc., of Houston, Texas. The streptomycetes are believed to protect the plants by producing antifungal antibiotics or enzymes that destroy fungal cell walls. Don Crawford's group at the University of Idaho found the *S. lydicus* strain to be doubly beneficial to peas when it grows in their rhizosphere. As well as killing fungal pathogens, it stimulates plant growth by providing iron in a usable form.[8]

In other symbioses between *Streptomyces* and plants, aerial tissues are colonized, such as the leaves of an Australian plant called snakevine. The use of this plant by aboriginal communities to disinfect wounds and promote their healing may depend on the production of antibiotics. These compounds were discovered by Gary Strobel of Montana State University and named munumbicins after an aboriginal leader, R. Munumbi Miller, who pointed the plant out to him.[9] The implication is that the antibiotic produced by its *Streptomyces* inhabitant might protect the plant. Again, we may ask whether there is specific recognition of a suitable host plant by *Streptomyces* spores, causing them to germinate when the host is sensed.

The Vegetative Mycelium

In a typical rod-shaped bacterium, the cells elongate by making new wall material along the length of the rod, then divide in the middle to make two daughters that repeat the process. The population increases from one to two, then four, eight, and so on exponentially, as long as there are enough nutrients. *Streptomyces* spores germinate by swelling and producing a fine tubular projection, sometimes two or even three, giving rise to vegetative hyphae that grow only at the tips. This is shown in Figure 6.6, in which vancomycin coupled to a fluorescent stain has bound to places where new cell wall is being made. The tips would have to extend faster and faster to accommodate an exponential increase in cell mass, and this could be sustained only up to a certain speed. Each hypha therefore produces a side branch to give two tips, and so on repeatedly, resulting in an approximately exponential increase in the bulk of the colony, even though it remains as an interconnected series of hyphae. On a good source of nutrients, the result is the mat of hyphae that we call the substrate or vegetative mycelium.

The mycelial habit of *Streptomyces*, like that independently evolved by molds, is an adaptation to colonizing solid substrates, such as the remains of plants and animals in the soil. The hyphae burrow into the food source and produce digestive enzymes to break it down. *Streptomyces* hyphae can also cross air spaces in the soil, in the same way as (but on a much smaller scale than) the dry rot fungus extends for long distances from a damp spot in a house to reach dry timber some distance away. In the soil, this gives *Streptomyces* an advantage over motile bacteria that thrive only in liquid films.

Although cross-walls are not made in *Streptomyces* hyphae every time the cell mass doubles, they are formed at intervals (Figure 6.6). In rod-shaped bacteria, many genes have been identified that encode proteins needed to make cross-walls. A key player is FtsZ, which is deposited as a ring inside the cell wall in positions where cross-walls will form. The ring grows inward like the diaphragm of a camera lens

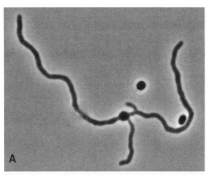

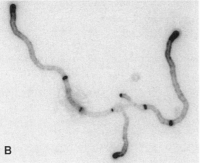

Figure 6.6. A spore of *Streptomyces coelicolor* has produced three germ tubes that are developing into young vegetative mycelium (two ungerminated spores are also seen). (A) Unstained. (B) Stained with a fluorescent dye coupled to vancomycin, which binds to hyphal tips and developing cross walls, where new wall material is being made. (Courtesy of Klas Flärdh, University of Lund.)

closing, attracting other proteins that form the new cross-wall, which cuts the parent into two new cells. Because unicellular bacteria can propagate only by making daughter cells, FtsZ is essential for their viability, but Joe McCormick in Richard (Rich) Losick's laboratory at Harvard found that a mutant of *S. coelicolor* lacking FtsZ was viable, even though it made no cross-walls in any part of the colony.[10] It could not make spores but was otherwise remarkably normal in appearance, although the colony is presumably a single giant cell. The mutant could be propagated by breaking the mycelium into pieces, many of which died, presumably because the cell contents oozed out of the broken ends, but a few remained viable, perhaps when the jelly-like cytoplasm sealed the ends in time. In the wild type, all except very short hyphal fragments are viable, pointing to the importance of cross-walls in protecting the colony when part of it is damaged.

Making New Chromosomes

As the mycelium grows, the chromosomes must replicate and populate the new hyphal compartments. These are cylinders with diameters of about 0.5 micrometers (μm; 1 μm = 1/1000 of a millimeter) and lengths of about 10 μm (Figure 6.6). The length of an unfolded chromosome, almost 2.2 mm, is 200 times the length of the cylinder and more than 4000 times its diameter, so the chromosome has to be folded very tightly to fit into the mycelium, and each compartment may contain a dozen or more copies of the chromosome. They must avoid becoming entangled and at the same time make new copies to provide chromosomes for the side branches as they form. Meanwhile, appropriate segments of the DNA are transcribed into messenger RNA to express the genes. This is an amazing feat of microengineering.

The biochemistry of chromosome replication is similar in all bacteria, but linearity of the *Streptomyces* chromosome poses special problems. A group led by Jolanta Zakrzenska-Czerwinska in Wrocław, Poland, has made a detailed study of *S. coelicolor* chromosome replication.[11] Near the center of the chromosome is a region called the origin of replication, *oriC* (C is for chromosome), where DNA synthesis begins. As in *E. coli*, in which it was first studied, the *oriC* region contains a gene encoding an initiator protein called DnaA, which is needed to start DNA replication by binding to a special nine-base-pair sequence. In *S. coelicolor*, there are 19 copies of this sequence in just 800 base pairs around *oriC*, providing an extended target for DnaA.

Replication begins when several molecules of DnaA bind to each target sequence in the *oriC* region. The DnaA molecules then stick together, making the DNA bunch up and partially unwind. This allows a second protein, a helicase, to insert into the DNA double helix and continue the unwinding. Then a complex of other proteins begins to make new DNA strands that match the preexisting strands through the rules of base pairing, with A opposite T and G opposite C. Two replication forks are established and travel outward from *oriC*, one along each chromosome arm. Extrapolating from *E. coli*, 25 to 30 different proteins are probably needed for the whole process of chromosome replication, which runs at about 700 base pairs per second, so the chromosome takes 100 minutes to copy. And the process is so accurate that a wrong base is incorporated only once every billion bases—100 entire *S. coelicolor* chromosomes.

One of the special properties of the enzyme that synthesizes DNA, the DNA polymerase, is that it works only in one direction, 5´ to 3´ (see the structure of DNA in Figure 4.7). This means that a replication fork cannot progress in what might seem the logical way, with the two new strands growing in the direction in which the replication fork is traveling. One strand, called the continuous strand, can do this, but the other has to be constantly reinitiated and extended back to join onto the segment of new DNA that has just been made (Figure 6.7). These segments of the discontinuous strand, called Okazaki fragments after their discoverer, are about 2000 bases long, so a new fragment has to start every 2 or 3 seconds, thousands of times as the whole chromosome replicates.

When a replication fork nears an end of the *Streptomyces* chromosome, there is a potential problem: while the continuous daughter strand can form right to the end, the discontinuous strand cannot. This is because of another property of the DNA polymerase:

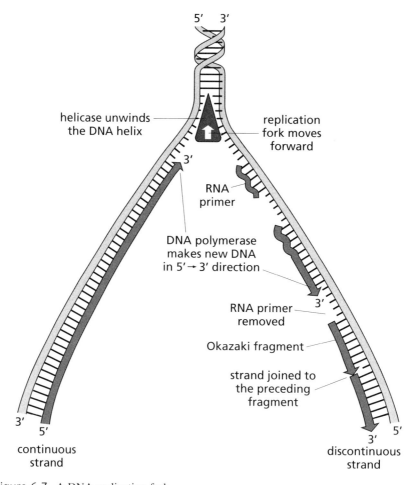

Figure 6.7. A DNA replication fork.

it can only start making new DNA by extending an existing strand. This "primer" is made by another enzyme and is a short stretch of RNA, not DNA. After use, the RNA primer is digested away, leaving a gap, which is filled by the next Okazaki fragment as it comes along (Figure 6.7). But when the primer for the last fragment was removed, a piece at the end of the discontinuous strand would be missing, so genetic information would be lost to the next generation, and so on progressively.

Organisms have solved this universal "end problem" in various ways. Most bacteria have adopted a very simple solution: their chromosomes are circular, so the replication fork running in one direction meets the one coming the other way and the new strands join up. Eukaryotes, with their linear chromosomes, have evolved a sophisticated mechanism in which the ends of the chromosomes consist of many copies of the same sequence. The chromosomes progressively lose DNA from the ends as they replicate, but a very unusual enzyme adds segments back to stop the chromosome's getting shorter. The process is not perfect, and one factor in human aging seems to be a gradual loss of the repeated copies at the ends of the chromosomes in some of our cells, so that they eventually lose genes. *Streptomyces* adopts a unique solution to the end problem. As Stan Cohen's group at Stanford discovered, a molecule of a special protein that is chemically bonded to the free 5´ end of the chromosome primes DNA synthesis, allowing the DNA polymerase to make the terminal Okazaki fragment from the very end of the chromosome without the need for an RNA primer.[12]

If most bacteria have solved the end problem by the apparently simple expedient of having a circular chromosome, why does *Streptomyces* have a linear chromosome? The question becomes even more intriguing if one considers that *Streptomyces* chromosomes occasionally become circular in the laboratory, and presumably in the soil, as a result of spontaneous loss of the ends and rejoining of the broken molecules; the resulting circular chromosomes can then replicate just like those of other bacteria. There must be some selective advantage in having linear chromosomes to offset the problems they pose for replication. Perhaps it has to do with the fact that, with so many copies of the chromosome in each compartment of the mycelium, circular chromosomes would become interlinked during DNA replication and recombination and then could not separate properly to be distributed to hyphal branches and spores as they form in the multicellular colony. This is obviously not fatal, otherwise mutants with circular chromosomes could not survive at all, and bacteria have a mechanism to separate linked DNA circles, but it might slow down colony growth just enough to be a long-term disadvantage.

Distributing the Chromosomes

How daughter chromosomes are distributed to daughter cells has intrigued researchers from the beginning of bacterial genetics. In unicellular bacteria, chromosome replication is strictly coupled with cell division: when a cross-wall forms in the center of the cell, one chromosome copy ends up in each new cell, in a process called chromosome partitioning. Cross-wall formation is not directly coupled with chromosome replication in *Streptomyces*, which has multiple copies of the chromosome in each hyphal compartment, but special mechanisms must operate to ensure that hyphal side branches and spores acquire copies.

Very close to the *S. coelicolor oriC* is a gene that encodes a partitioning protein called ParB, recognized by its similarity to such genes in other bacteria. If this gene is mutated, problems arise, causing an upset in the normally precise distribution of one chromosome into each spore. ParB binds to a special 16-base-pair sequence found in 24 copies in a 400-base-pair segment centered on *oriC*. Dagmara Jakimowicz, working as a postdoctoral fellow with Keith Chater at the John Innes Centre as part of a collaboration with the Wrocław group (Figure 6.8), showed that this concentration of sites is the target to which ParB binds, making a huge complex of protein and folded DNA around *oriC*.[13] In the aerial mycelium Dagmara saw these complexes regularly spaced at positions where the spore genomes would end up, suggesting that they are involved in the correct spacing of chromosomes to be included in the spores. In the vegetative mycelium the complexes were mostly just behind hyphal tips, suggesting that, as new tips form to initiate side branches, they are coupled to *oriC* so that the DNA is dragged into the side branch as it develops.

As well as having to move during vegetative growth, chromosomes are transferred between hyphae by mating, under the influence of plasmids. The paradigm for DNA transfer during bacterial mating is the F plasmid in *E. coli* (Figure 6.9). F carries at

Figure 6.8. Jolanta Zakrzewska-Czerwinska (left) and Dagmara Jakimowicz in the Ludwik Hirszfeld Institute of Immunology, Wrocław University, May 2005. (Courtesy of Jolanta Zakrzewska-Czerwinska.)

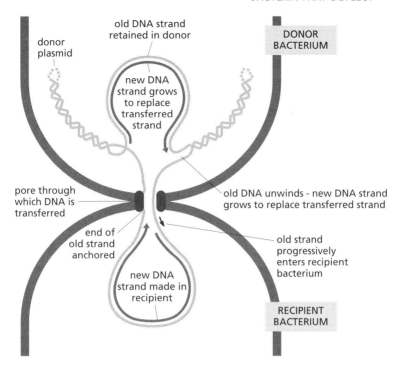

Figure 6.9. Transfer of plasmid DNA from an *Escherichia coli* donor to a recipient.

least 35 genes needed for transfer, including 15 genes encoding proteins that build thread-like "pili" on the donor cell surface. They make contact with the recipient, draw it to the donor and establish a narrow channel between the cells. Plasmid-encoded enzymes then make a single-strand break in the DNA at a unique site on the plasmid, and another protein guides one of the resulting free ends through the mating union into the recipient, where the transferred DNA is converted back to a double strand, while the donor synthesizes a new strand to replace the one transferred. The end result is a complete plasmid in both bacteria. Transfer of chromosomal DNA occurs as an extension of plasmid DNA transfer, when F is integrated into the host chromosome, as described in Chapter 3 (Figure 3.13).

We now know that DNA transfer occurs very differently in *Streptomyces*. First of all, there is no mating structure equivalent to the pili. Transfer probably occurs when hyphae coalesce at the tips, where the wall is growing by breaking linkages between the components and inserting new material. In marked contrast to the *E. coli* paradigm, *Streptomyces* plasmids typically carry a single gene for DNA transfer, and the protein it encodes is located near hyphal tips, as shown by recent work in Günther Muth's group at the University of Tübingen, Germany,[14] strongly supporting the idea of plasmid transfer by tip fusion. Second, the DNA almost certainly remains double-stranded during transfer. Circumstantial evidence for this comes from the similarity between the *Streptomyces* transfer proteins and a protein called SpoIIIE, which Jeff Errington's group in

Oxford showed acts as a DNA motor to drive a double-stranded chromosome into the developing *Bacillus subtilis* spore.[15] Günther Muth's group has identified a specific double-stranded sequence on a *Streptomyces* plasmid to which the transfer protein binds, an action that is necessary for plasmid transfer. And the group of Michel Guérineau and Jean-Luc Pernodet at the University of Paris-Sud obtained experimental evidence that a plasmid in *Streptomyces ambofaciens* is transferred as double-stranded DNA.[16]

A special problem faced by *Streptomyces* plasmids, after transfer by mating, is to spread throughout the multicellular colony of the recipient, migrating from the point of transfer, compartment by compartment. Migration is promoted by so-called "spread" genes, which are unique to *Streptomyces* plasmids and were detected because they are responsible for the pocks used to detect artificially induced transformation by plasmid DNA (Chapter 4). When plasmid-carrying spores germinate in a plasmid-free lawn, pocks up to a few millimeters across are seen, like those in Figure 4.9. The pocks represent the extent of the mycelium into which plasmid copies have migrated, so spreading may involve a "journey" of a couple of millimeters, a huge distance in molecular terms. It is still a mysterious process.

There remains the question of how *Streptomyces* plasmids interact with chromosomal DNA to promote its transfer. We know that SCP1 can occasionally be inserted into the *S. coelicolor* chromosome by crossing-over between IS sequences, as with the F plasmid in *E. coli*, losing its ends in the process so that the chromosome is not broken at the point of integration.[17] Transfer of the plasmid then drags the chromosome into the recipient, again presumably in double-stranded form. How autonomous plasmids transfer the chromosome is still mysterious. Circular plasmids have not been found to integrate stably into the chromosome via some type of insertion sequences, as F does, but may perhaps undergo transient interactions with the chromosome. In contrast, Carton Chen suggested that a linear plasmid such as SCP1 can transfer the linear chromosome end-first into the recipient by a different mechanism. To interact, the ends of the two molecules were postulated to have a unique three-dimensional structure determined by a series of repeated DNA sequences that fold back on themselves (Figure 6.10). The resulting structure would be recognized by special binding proteins that in turn stick to each other to hold the two double-stranded DNA molecules together.[18]

The idea that chromosome ends have a tendency to bind together was supported by a nice experiment by Melody Yang in Rich Losick's laboratory at Harvard,[19] who found that the two ends of the linear *S. coelicolor* chromosome normally associate with each other in the vegetative mycelium. She deduced this by using DNA probes tagged with fluorescent dyes to light up specific regions of the chromosome. When one of two differently tagged probes hybridized to a sequence near one of the chromosome ends, and the other to the opposite end, the colored spots of light emitted by the two dyes were almost all close together, whereas when one probe hybridized near *oriC* and the other near a chromosome end, the spots were randomly distributed in the hyphae.

These recent findings show that mating in *Streptomyces* occurs by very different means from those in *E. coli* and other bacteria, including all the gram-negative species and non-actinomycete gram-positive species so far investigated. From a personal viewpoint it is exciting that one of my reasons to carry on working on *Streptomyces* genetics after they were shown to be bacteria—to discover novel genetic behavior

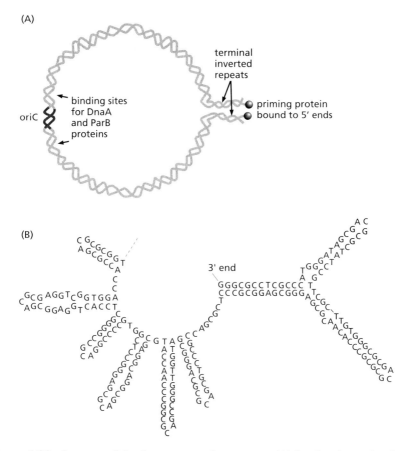

Figure 6.10. Structure of the *Streptomyces* chromosome. (A) Landmarks on the chromosome. (B) Structure that could be formed by pairing between complementary DNA sequences within a single-stranded 3' end region.

(Chapter 3)—has been vindicated.[20] It seems remarkable that plasmids are essential for mating in *Streptomyces*, as in other bacteria, but that the mechanisms they control are so different. There is a lot more to discover about these special processes, but the next year or two should see many new insights.

The Rise of the Aerial Mycelium

After growth of the vegetative mycelium, the next and perhaps the most important decision a *Streptomyces* colony must make during its life cycle is to go into survival and dispersal mode. The first step on this path to sporulation is to initiate the aerial mycelium.

The developing aerial hyphae must leave the moist substrate on which the vegetative mycelium has been feeding and enter the air. This presents a physical challenge, because they must break through a water-air interface that has a high surface tension, which must be reduced if the hyphae are to emerge. Detergent-like proteins secreted by the hyphae mediate this change. The chaplins probably play a role, and another player is a very small protein called SapB, which was studied by Joanne Willey, first at Harvard and then at Hofstra University on Long Island, New York. In a collaboration with Justin Nodwell at McMaster University in Hamilton, Ontario, and Mark Buttner's group at John Innes, SapB was found to resemble a class of antibiotics called lantibiotics,[21] raising the possibility that SapB might play a dual role: helping the aerial hyphae to enter the air and, if it remains on the surface of the spores, protecting them from attack by other microbes.

In a striking example of the parallel evolution of filamentous bacteria and fungi, Joanne and her Dutch collaborator Han Wösten found that some *S. coelicolor* mutants that are unable to form aerial mycelium and are devoid of SapB and the chaplins regained the ability to erect aerial hyphae when a fungal protein was added to them. This "hydrophobin" evidently plays a similar role in the fungus as SapB and the chaplins do in *Streptomyces*.[22] The proteins are unrelated in amino acid sequence, and so in evolutionary origin, yet they have equivalent physical properties.

Nutrient Recycling

Where do the nutrients come from to make the aerial mycelium and spores? A Spanish group at the University of Oviedo, led by Carlos Hardisson, showed by radioactive labeling that material from the vegetative mycelium is recycled and ends up in the spores. At first it was assumed that the vegetative hyphae were simply degraded in an unordered way, but the Oviedo group recently reinterpreted what is going on in the central regions of the *Streptomyces* colony during this process. They likened it to programmed cell death, a process that has huge implications in the development of animals and of cancers.[23] As the name implies, this is not a random cell degradation but a disassembly of the contents of cells once they have been targeted for destruction. An analogy would be the systematic taking down of a building by a team of carpenters, plumbers, and masons, rather than simply sending in a crane with a wrecking ball.

As soon as it was realized that the older parts of the *Streptomyces* colony were being broken down by digestive enzymes, the question arose: What prevents them from destroying the young hyphae still foraging at the colony edge? The complete answer must be complex, but a group at Seoul National University, led by Kye-Joon Lee, has provided a partial explanation, involving three proteins.[24] This group proposed that an enzyme very similar to the trypsin in human stomachs serves to break down proteins in the older mycelium. Young hyphae in the outer regions of the colony are protected from digestion by a small, hydrophobic protein called leupeptin that inhibits the trypsin-like enzyme. The third player is a leupeptin-inhibiting enzyme that reverses the protective action of the leupeptin and allows the trypsin to act as the hyphae age. As in the development of other organisms, the question is begged as to what mechanism establishes the gradients of the key molecules (here the leupeptin-inhibiting enzyme), but identifying some of the players is a step toward understanding the game plan.

Developmental Switches

Except for a few special situations, it is a golden rule that all members of a microbial colony, and all the different cells in the tissues of a eukaryote, have the same genetic potential. This was strikingly confirmed by the cloning even of mammals such as Dolly the sheep from specialized adult cells. All the cells in the microbial colony or the multicellular organism express a certain subset of their genes that are responsible for housekeeping functions, to build the cell and to control its basic mechanisms and metabolism, but when cells perform specialized roles, such as the cells in the mammalian pancreas dedicated to making insulin or the *Streptomyces* spore as a resting and dispersive stage in the life cycle, they switch on particular genes to give rise to the specialized properties of the different cell types.

François Jacob and Jacques Monod at the Pasteur Institute in Paris described the first example of such a gene switch in molecular terms around 1960. *E. coli* utilizes lactose, a disaccharide made of the simple sugars glucose and galactose joined together, using an enzyme called beta-galactosidase that breaks the bond between the two sugars. It would be a waste of energy for the bacterium to produce the enzyme all the time, so a genetic switch ensures that the beta-galactosidase gene is transcribed only in the presence of lactose. The French team defined a special DNA sequence (the promoter), upstream of the beta-galactosidase gene, with which the RNA polymerase interacts. A repressor that binds to another sequence (the operator) blocks access of RNA polymerase to the promoter. However, when a derivative of lactose, the "inducer" of the genes, is available, it binds to the repressor, causing it to change its shape, lose its affinity for the operator, and allow transcription to proceed.[25]

The repressor postulated by Jacob and Monod from their genetic experiments was the first example of a class of proteins called transcription factors. While the lactose transcription factor acts negatively, blocking the action of the RNA polymerase, examples of the opposite behavior were soon found. AraC, an activator of the genes for arabinose utilization in *E. coli*, was discovered by Ellis Englesberg at the University of Pittsburgh in 1965.[26]

The annotators of the *S. coelicolor* sequence predicted that 965 of the 7825 genes would encode transcription factors, because the proteins contain characteristic amino acid motifs that bind to DNA. This is the highest proportion of regulatory genes in any bacterial genome sequenced so far. Some of them control developmental switching to initiate the aerial mycelium and its metamorphosis into spores.

Bacteria also use another mechanism to control transcription of appropriate groups of genes. The bacterial RNA polymerase consists of two components: the core, made from several proteins, and an extra protein called the sigma factor. The core can synthesize RNA on a DNA template, but it is not specific for particular DNA sequences: it is the sigma factor, when it joins with the core, that provides the specificity to recognize appropriate promoter sequences and start transcribing only from them. Bacterial promoters have been studied in enormous molecular detail since Jacob and Monod defined them genetically. They typically contain two six-base-pair DNA sequences, called the –10 and –35 sequences because they are centered on positions 10 base pairs and 35 base pairs upstream of the point where transcription starts. These are the sequences that make specific contacts with the sigma factor. All bacteria have one sigma factor that recognizes the promoters of the housekeeping genes; usually, others transcribe different groups of genes, for developmental or other adaptive responses, by recognizing alternative –10 and –35 promoter sequences. It was Jan Westpheling, while she was a postdoctoral fellow with Rich Losick at Harvard, who was the first to show that *Streptomyces* uses such a strategy.[27]

Developmental Mutants

In the study of development, mutants that have lost the ability to complete the process allow us to highlight different steps in the whole event. *Streptomyces* is good for studying development, because such mutants can be identified simply by looking at the hundreds of colonies growing after a mutagenic treatment. They survive without making spores because they can be propagated indefinitely from mycelial fragments, whereas mutations in developmental pathways are lethal in many organisms.

The *S. coelicolor* mutants are of two main kinds. One class cannot produce aerial mycelium and lack the hairy layer that covers normal colonies. They appear smooth and shiny, like the pate of a bald man, so we call them bald mutants and give them and the genes they identify a standard three-letter genetic symbol, *bld*. The colony on the right in Figure 6.11 is a *bld* mutant, compared with the wild type on the left. Mutants of the second class make aerial hyphae but not spores. *S. coelicolor* spores are pale gray-brown, so colonies that fail to develop right through to spore maturation appear white as the aerial mycelium scatters light like snow. The mutants are therefore called *whi*.

We picked the first *bld* mutants in the mid-1960s while I was still in Glasgow, and Mike Merrick isolated others in the early 1970s as a postdoctoral fellow in Norwich. More than 20 *bld* genes have now been identified, and new ones are being found in laboratories around the world. When Liz Lawlor (Figure 6.12), a PhD student working with Keith Chater (Figure 6.13), sequenced the first *bld* gene in 1987, she got a big surprise: the DNA did not appear to encode a protein. Howard Baylis, a graduate

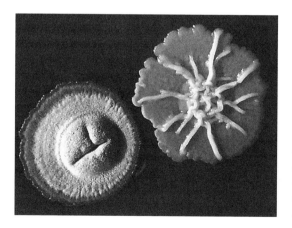

Figure 6.11. A colony of a *bldA* mutant of *S. coelicolor* (right) compared with the wild type (left). There is no good explanation for the ridges on the surface of the *bldA* colony. (Courtesy of Tobias Kieser, John Innes Centre.)

Figure 6.12. Elizabeth (Liz) Lawlor at the John Innes Centre, 1987, with the DNA sequence that proved the *bldA* gene of *Streptomyces coelicolor* to be a tRNA. (Courtesy of Tobias Kieser.)

student working with Mervyn Bibb (Figure 6.13), had become used to looking at chromosomal sequences for the transfer RNAs (tRNAs) of bacteria, and he recognized that the *bldA* gene specified one of them.[28] The tRNAs are essential for translating the messenger RNA into protein on the ribosomes, according to the genetic code. Each tRNA has a triplet of bases, called an anticodon, that is complementary to one of the messenger RNA codons. A specific enzyme loads each tRNA with the amino acid corresponding to that codon, and the tRNA then aligns the amino acid on the messenger RNA, ready to be joined to the amino acid corresponding to the previous codon. The *bldA* tRNA is for one of the six leucine codons, TTA in the DNA code.

Liz could immediately deduce something remarkable—no TTA codons could occur in any *S. coelicolor* gene that is essential for vegetative growth, because the *bldA* mutant grows normally up to the stage at which aerial mycelium would be produced by the wild type. Almost all proteins contain leucine, but other codons must be used to specify it in the proteins needed for vegetative growth. The *bldA* gene operates a novel kind of developmental switch, in which tRNA for the TTA codon appears only as aerial mycelium is due to develop, ready to translate the messenger RNAs of genes needed for aerial hyphal initiation. We do not yet know what controls this key timing event, but we do know the immediate target of the *bldA* developmental switch.

The genome sequence shows that TTA is used for only 4 of every 1000 leucines, but, even so, 145 *S. coelicolor* genes contain a TTA codon (or occasionally two). Eriko Takano and Tao Meifang at the John Innes Centre and Kien Nguyen in Charles Thompson's group in Basel, Switzerland, identified the *bldA* target as one of these TTA-containing genes.[29] It turned out to be *bldH*, which encodes a transcription factor belonging to a large class exemplified by the classical AraC activator of *E. coli*. Ex-

Figure 6.13. Keith Chater (left) and Mervyn Bibb at the John Innes Centre, September 1998. (Courtesy of Tobias Kieser.)

pression of *bldA* leads directly to translation of the *bldH* transcript, and this triggers a cascade of several steps, still to be worked out in detail, leading to aerial hyphae.

The Path to Sporulation

Once the *bld* genes have initiated the aerial hyphae, a lot of decisions still lie ahead. After elongating for a while, the aerial hyphae stop growing and the apical compartment is delimited by a cross-wall at the base. The hyphae coil, and the apical compartment is subdivided by special cross-walls into 50 to 100 short cylindrical segments, each of which rounds up and becomes a spore. Back in 1969, Keith Chater started a study of 50 *whi* mutants, which Helen Kieser and I had isolated in Glasgow, by classifying them into nine groups according to their morphology and by genetic mapping, labeling the groups *A* through *I* in order of the genes' positions on the chromosome.

It is *whiG* that controls the switch from elongation to sporulation: its disruption causes the hyphae to continue growing, as seen in Figure 6.14. When Carmen Mendez, a Spanish postdoctoral fellow working with Keith, cloned *whiG* in 1987 and it was sequenced in Keith's group a couple of years later, they got a nice surprise. The gene encodes an RNA polymerase sigma factor, clearly implicating sigma factors in the developmental cascade.[30]

The idea that developmental switches in bacteria are controlled by specific sigma factors stemmed from work with *Bacillus subtilis* by a married couple, Rich Losick and Janice Pero, at Harvard, starting around 1980. Rich was one of the pioneers of sigma factors and has made a career-long study of the developmental cycle of *B. subtilis*. This bacterium is very different from *S. coelicolor*; it grows vegetatively by simple elongation of the rod-shaped cells, followed by their splitting into two identical daughters. Their developmental repertoire is to produce highly resistant spores, one inside each cell, when the going gets hard. They are some of nature's toughest structures, able to withstand boiling in water for a long time and not killed rapidly

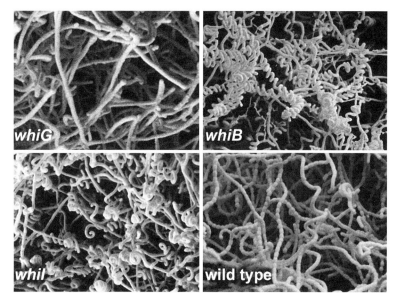

Figure 6.14. Scanning electron micrographs of *whi* mutants of *Streptomyces coelicolor* compared with the wild type. The mutants are interrupted at various steps in normal differentiation, during which the aerial hyphae stop elongating and undergo coiling and segmentation into spore compartments. In *whiG* the aerial hyphae continue to elongate and show no signs of coiling or segmentation into spore compartments; in *whiB* the aerial hyphae continue to elongate but undergo extensive coiling and no segmentation; in *whiI* the aerial hyphae stop growing and become coiled but fail to undergo segmentation. (Courtesy of Jamie Ryding and Kim Findlay, John Innes Centre.)

until the temperature rises to about 120°C. The bacterium that causes anthrax, *Bacillus anthracis*, makes such spores; hence, its potential as a biological weapon: the spores could survive long enough to be distributed widely in a terrorist attack. *B. subtilis* spore development involves a cascade of four sporulation-specific sigma factors that switch on different groups of genes progressively as the spore develops.[31]

A second *S. coelicolor whi* gene has turned out to encode another sigma factor, and most of the others specify transcription factors of various classes, which control different events during the metamorphosis into spores.[32] For example, in a *whiB* mutant, the hyphae fail to stop elongating, but the next obvious morphological event, coiling of the hyphae into spirals, goes ahead and the mutant shows a mass of unsegmented, corkscrew-like mycelium (Figure 6.14). In a *whiI* mutant, in contrast, the aerial hyphae stop elongating and become coiled normally but again fail to form cross-walls (Figure 6.14). Many other *whi* mutants are now being studied in Europe, North America, China, and Japan.

With the complete genome sequence of *S. coelicolor* available, why can we not make an inventory of all the genes involved in aerial mycelium and spore development just by scanning the sequence? The answer is that we cannot predict accurately enough what kinds of proteins might be needed. This has to be determined experimentally through the study of mutants. However, instead of just isolating *bld* and *whi* mutants by the traditional methods of genetics, in which random mutants identify the genes of interest, we can now guess which genes might be involved, then systematically inactivate the genes and examine the consequences. We shall come to these "functional genomics" techniques in Chapter 9.

It is significant that almost all of the *bld* and *whi* genes identified in *S. coelicolor* by mutants with incomplete development play regulatory roles rather than coding for enzymes or structural proteins: the only clear exception is *whiE,* which simply encodes the biosynthetic enzymes that make the spore-associated pigment. The situation is similar for development in many other organisms. The explanation is probably that the aerial mycelium and spores of *Streptomyces* are constructed largely using components and enzymes essential for building the vegetative mycelium. This means that mutations in the genes encoding these components would be lethal. Instead, the regulators of gene expression—a tRNA, sporulation-dedicated sigma factors, and specific transcription factors of various classes—which marshal the same components as are used to build vegetative cells, but in new ways, can be mutated without killing the organism.

Tissue-Specific Enzymes

In plants and animals, similar biochemical reactions are often carried out in different parts of the organism. Sometimes the same enzymes, encoded by the same genes, are used; in other cases, different genes encode almost identical tissue-specific variants of the enzyme. Does this happen in *S. coelicolor* with its different tissues, unique among bacteria? The answer is "yes—sometimes." The best example comes from work by Keith Chater's group in *S. coelicolor* on glycogen, building on an earlier discovery by Carlos Hardisson's team in Oviedo in a different *Streptomyces* species.

Glycogen, a starch-like polymer of glucose, is the major store of energy in the liver and also plays this role in many bacteria. In *S. coelicolor* colonies, glycogen granules appear in two different tissues: the mature vegetative mycelium and the immature spores. The glycogen is then broken down again to supply energy and building blocks for developmental processes. In the vegetative mycelium this helps to power aerial mycelium development, and in the young spores the sugar supplies carbon and energy for them to mature. The interesting thing is that two sets of similar enzymes, encoded by different genes, control some of the steps in the synthesis and breakdown of glycogen in the two tissues, just as in a plant or animal.[33] Many more examples of such tissue-specific forms of the same proteins will doubtless be identified, as was hinted by the striking finding of many duplicated gene sets in the genome. For example, it is intriguing that each of the mysterious gas vacuole genes is actually present in two copies in *S. coelicolor*, not just one, and *S. avermitilis* even has three sets. Almost certainly, many of these apparent duplications will be found to encode tissue-specific forms of the same proteins. Their further study will throw more light on the events that occur in the various morphologically different phases of the developmental cycle that set *Streptomyces* apart from non-actinomycete bacteria. The other special features of *Streptomyces* are physiological: their capacity to recognize a multitude of chemical and other stresses and opportunities in the soil and respond appropriately, including by making antibiotics. I describe these responses in the next chapter.

7

The Switch to Antibiotic Production

From a human perspective, the most important biochemical feature of the streptomycetes is their capacity to make antibiotics. From the organism's point of view, this is an adaptation to competition in the soil, with its teaming populations of microbes and small invertebrates of all kinds. The regulation of antibiotic production is a very sophisticated process, optimized over eons of natural selection to maximize the selective advantage of the organism.

Competition among microbes largely boils down to a struggle for food. In the *Streptomyces* life cycle, a vegetative mycelium forages for nutrients and assembles them into cellular structures and storage compounds. Later, this "biomass" is recycled to build the aerial mycelium and eventually the spores for reproduction and dispersal (Chapter 6). Antibiotics are made at the transition between these two phases of the life cycle. Their production is a response to nutritional stresses sensed by the colony as it communicates with its environment, exporting enzymes, importing nutrients, and monitoring their concentrations by relays of sensory and regulatory signals that lead to the expression of appropriate suites of genes. All microbes have such systems, but they are particularly comprehensive in *Streptomyces*, adapting the organisms superbly to life in the soil. In this chapter, I consider some of this physiological interplay between the organisms and their environment and then go on to describe how antibiotic biosynthesis itself is initiated and regulated.

Cell Membranes and Their Channels

Biological membranes are the interface between the inside of the cell and its environment. They play a challenging role in enclosing the cellular contents, including the hundreds of small molecules that make up the organism's metabolism, but allowing molecules to enter or leave as appropriate. The membranes consist of a double layer of lipid molecules that have a head with an affinity for water and two long, oily tails. Each layer of the membrane has closely packed heads at the surface, with the tails extending inward, where they meet the tails of the other layer pointing toward them (Figure 7.1).

Such a simple membrane would allow through molecules with an affinity for fats—so-called lipophilic or hydrophobic molecules—because they could slide past the oily tails in the membrane, but water, as well as hydrophilic molecules that are soluble in water but not in fats, would be excluded. The membrane therefore has many transporter proteins inserted into it, each allowing particular kinds of molecules to diffuse or be pumped through the membrane. The *Streptomyces coelicolor* genome sequence contains about 450 genes predicted to encode transporter proteins, compared with fewer than 300 for *Escherichia coli* and 150 for *Mycobacterium tuberculosis*. We cannot yet deduce all the molecules they carry, but the finding immediately tells us that *S. coelicolor* can import or export a huge range of molecules.

Small molecules such as inorganic ions, sugars, amino acids, and antibiotics, as well as water itself, cross cell membranes through water-containing channels. One

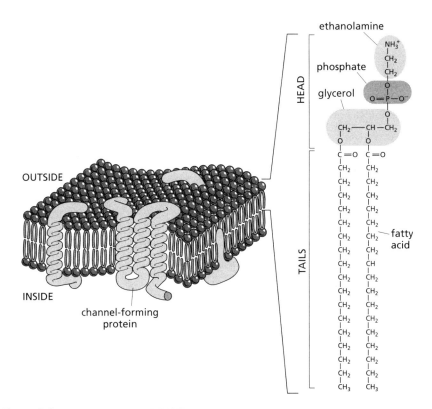

Figure 7.1. A cell membrane lipid bilayer.

of the simplest is the tetracycline resistance pump (we saw in Chapter 2 that pumping out the antibiotic confers resistance to tetracycline). Each channel consists of a protein with ends made up mainly of hydrophilic amino acids. One end protrudes into the outside medium, and the other extends into the cytoplasm on the other side of the membrane. Between the two ends, the protein crosses the membrane no fewer than seven times. Each of the seven transmembrane segments consists of about 20 amino acids that form a rigid, rod-like structure, called an alpha helix, that is just long enough to cross the membrane. These amino acids are mainly hydrophobic so they interact well with the oily tails of the membrane. Between successive alpha helices are runs of mainly hydrophilic amino acids that allow the protein to loop back on itself to bring the seven helices parallel to each other in a bundle. The whole structure forms a tube across the membrane through which particular molecules can pass, depending on affinities between the transported molecule and the inside of the channel. One of these channel-forming proteins is seen sliced through in Figure 7.1.

Transport of tetracycline across the membrane is driven by an electrical charge difference between the inside of the cell and the outside medium: cells contain an excess of negatively charged hydroxyl groups (OH^-), and the medium has an excess of positively charged protons (H^+). Tetracycline is positively charged, and export of each molecule is driven by uptake of a compensating proton. Another type of channel, called ABC transporters (for *A*TP-*b*inding *c*assette), use adenosine triphosphate (ATP), the cell's chief energy-storing fuel, to drive the transport process.[1] A component of the transporter binds ATP through a characteristic amino acid sequence and lies inside the cell, attached to membrane-spanning helices that form a channel across the membrane similar to the tetracycline transporter, as well as a component that binds to whatever compounds each transporter is adapted to recognize (Figure 7.2). When a substance is to be exported, the channel first opens to the cell's interior, and the

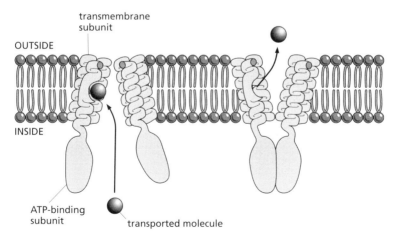

Figure 7.2. An ABC transporter opening to the inside of the cell (left) and to the outside (right). (The trans-membrane and ATP-binding functions of the transporter are here shown on a single protein, but often they are distributed over two or three different proteins that join together to build the pump.)

molecule to be transported diffuses into it, triggering the channel to change shape and open to the outside, releasing the substance into the medium. The channel uses energy from ATP to revert to its original shape. All organisms, from the simplest bacteria to the most complex eukaryotes, are well endowed with ABC transporters. Annotation of the *S. coelicolor* genome sequence revealed 140 genes that would encode ABC transporters, compared with 80 in *E. coli* and 32 in *M. tuberculosis*.

An especially important and universal transporter allows potassium ions into cells, where they are crucial in maintaining appropriate electrical properties, accentuated in excitable tissues such as those in the vertebrate heart and nervous system. It was the cloning of the *Streptomyces lividans* potassium transporter by Hildgund Schrempf's group at the University of Osnabrück that allowed Roderick MacKinnon at the Rockefeller University in New York to make enough of the protein to determine its three-dimensional structure and thereby deduce how the transporter works, winning him the Nobel Prize for Chemistry in 2003. The mammalian transporter had proved impossible to work with, but the *S. lividans* version was much more amenable.[2] Four molecules, each with two transmembrane helices, combine to form a funnel-shaped channel that opens to the outside of the cell. There is a vestibule that receives the ions to be transported from the medium, followed by a narrow pore. This is so selective that it can carry millions of potassium ions per second, but only one sodium ion gets through for every thousand potassium ions, because sodium is too small to bind properly to the pore and be moved inward along it. Larger ions such as calcium and magnesium cannot pass the sieve-like lumen. On the inner face of the membrane, the pore is closed by a "gate" where the transmembrane helices cross like the struts of a tepee. These helices separate in response to a signal sensed by the interior of the channel when potassium is needed.

Proteins are transported through membranes in a totally different way from ions and small molecules. Proteins destined for export are made as precursors that are converted to the mature protein during transport from the cell. At the start of the protein (the N-terminus, so-called because of a free amino-group, $-NH_2$, at this end) is a characteristic stretch of hydrophobic amino acids called a signal sequence that inserts itself into the membrane. This is followed by the amino acid sequence that will form the mature functional protein, ending at the C-terminus (a free carboxyl group, –COOH). The signal sequence is recognized by a secretion apparatus made up of several proteins that form a channel across the membrane. It extrudes the secreted protein through the membrane like a thread and snips the mature protein off the signal sequence once it is outside. As many as 819 proteins encoded by genes in the *S. coelicolor* genome sequence, more than 10 percent of the entire protein complement, are predicted to start with a signal sequence and so be exported. Many of them are involved in utilizing complex food sources, which need to be digested outside the cell because they are much too large to be imported intact.

Scavenging for Nutrients

Streptomyces gains its competitive advantage in the soil by obtaining nutrients from sources that many other microbes cannot exploit, including complex carbohydrates

such as starch, cellulose, and chitin. These compounds have to
appropriate genes for their assimilation expressed just when th
 Exported enzymes called amylases break down starch, whic
molecules linked in long chains. Amylases attack the free ends
off pairs of glucose molecules as a disaccharide sugar called m
ces colony makes a small amount of amylase all the time, and
maltose when starch is encountered. The colony could also respond to maltose released
from starch by a competitor. Either way, a little maltose diffuses into the cells and interacts with a repressor, displacing it from its binding site on the DNA and allowing transcription of genes for an exported amylase, together with a transporter for importing maltose into the cell and an intracellular enzyme for splitting it into two glucose molecules. Maltose thus acts as a surrogate inducer for the proteins needed to metabolize starch, which cannot cross the membrane and induce gene expression itself.[3]

Because plant storage organs such as potatoes and cereal grains consist largely of starch, *Streptomyces* will come across lots of it in the soil. Reflecting this, the *S. coelicolor* genome has three genes encoding amylases. But by far the most abundant polysaccharide on earth is cellulose, followed by chitin. Cellulose is the main component of plant cell walls, including those that build tree trunks, and chitin forms the exoskeletons of insects and crustaceans and the walls of fungal hyphae. Only a minority of *Streptomyces* species can live on cellulose, but almost all of them use chitin, perhaps because the sugars that form it contain nitrogen, so chitin is especially valuable as a source of both carbon and nitrogen. The *S. coelicolor* genome has 8 genes predicted to encode secreted cellulose-degrading enzymes and as many as 13 for chitin. Jan Westpheling (seen in Figure 8.16), by then at the University of Georgia in Athens, and her group made a detailed analysis of the intricacies of the promoter of a *Streptomyces* chitinase gene, which allowed it to respond with exquisite accuracy to the levels of glucose and chitin in the environment.[4] Glucose, as a preferentially utilized carbon source, repressed expression whereas chitin, presumably via a soluble breakdown product as in the case of starch, induced it.

It is harder for organisms to use cellulose and chitin than starch because they are made of chains of sugar units aggregated into crystalline microfibrils that are very resistant to breakdown. Organisms that can attack them recognize the target polymers with receptor proteins that bind to cellulose microfibrils of plant cell walls, or to chitin-containing fungal hyphae or invertebrate exoskeletons (Figure 7.3). Hildgund Schrempf (Figure 7.4) has studied chitin degradation by *Streptomyces*. Some of her electron micrographs suggest that *Streptomyces* might even be able to attack the walls of living fungi it encounters in the soil.

Hildgund's group have made their most detailed studies on a signal relay that acts as a detection device for cellulose.[5] One end of a cellulose-sensing protein is extracellular and binds strongly to crystalline cellulose. It is followed by segments that cross the hyphal wall and the cytoplasmic membrane and place the other end of the protein in the cytoplasm. Four molecules probably lie parallel to each other in a bundle; this would change shape when the cellulose-binding region encounters crystalline cellulose, causing the internal region to interact with a transcription factor that switches on the genes needed to degrade cellulose. The system resembles receptors involved in many responses in the human body.

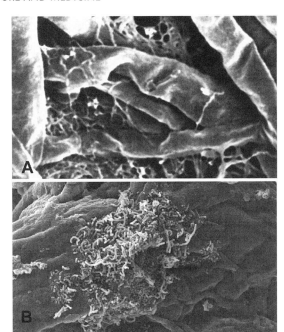

Figure 7.3. Scanning electron micrographs of *Streptomyces* binding to chitin. (A) Fungal hyphae (the large, collapsed tubular structures) with the very much smaller *Streptomyces* hyphae attached to their surface. (Courtesy of Hildgund Schrempf, University of Osnabrück). (B) A mass of *Streptomyces* (mostly already converted to spores) attached to a bag containing crab shells buried in soil for 5 months. (Courtesy of Martin Krsek and Elizabeth Wellington, University of Warwick.)

S. coelicolor obtains other difficult-to-utilize nutrients in different ways. All organisms need iron, because its unique atomic structure allows it to catalyze essential reactions that depend on the transfer of electrons, such as those in aerobic respiration. Iron is the most abundant metal on earth, but the strongly oxidizing atmosphere keeps almost all the iron in an oxidized state called FeIII that is virtually insoluble under physiological conditions. Microorganisms cope with this problem by secreting siderophores into the medium. These compounds have a very high affinity for FeIII, binding to it even when there is hardly any iron in solution by forming six hydrogen bonds to each iron atom from strategically placed oxygen or nitrogen atoms on the siderophore. Once it is loaded with iron, special transporters take the siderophore back into the cell, where the iron is converted to the FeII state and used.

It was expected that annotation of the *S. coelicolor* genome would reveal genes encoding enzymes for making siderophores, but it was a surprise to find three sets of genes for chemically different siderophores. Greg Challis of the University of Warwick has proposed an interesting explanation. One of the siderophores, desferrioxamine, is made by many streptomycetes; the Novartis company uses a strain of *Streptomyces pilosus* to produce this compound for treating potentially fatal iron overload (e.g., in thalassemia patients, who receive frequent blood transfusions). Within the cluster of genes for making desferrioxamine in the *S. coelicolor* genome, Greg found a gene encoding a transporter for desferrioxamine when loaded with iron. Many microbes that do not make this siderophore encode such transporter proteins and indulge in "siderophore piracy," snatching the iron–siderophore complex made available by organisms such as *Streptomyces*.[6] Perhaps this is why *S. coelicolor* makes

Figure 7.4. Hildgund Schrempf at the John Innes Centre, September 1998. (Courtesy of Tobias Kieser.)

a second siderophore that Greg has called coelichelin, which few competitors can take up. This would save *S. coelicolor* from iron starvation when its desferrioxamine is snatched. The third predicted siderophore of *S. coelicolor*, coelibactin, may play a different role. The gene cluster for its biosynthesis lacks a gene for an uptake protein, so coelibactin is unlikely to be used for iron scavenging by the producer. Instead, it could act as an unspecialized antibiotic by complexing free iron and starving competitors of this essential element.

Stress Management

During morphological development, a team of transcription factors and alternative RNA polymerase sigma factors mediates expression of appropriate sets of genes at different stages, as we saw in Chapter 6, and the same is true of the organism's physiological responses to chemical stresses in the soil.

Expression of vancomycin resistance is a good example of control by a specific transcription factor. The *S. coelicolor* genome sequence unexpectedly revealed a gene set closely resembling genes for vancomycin resistance in the actinomycetes that produce the antibiotic and in pathogens that have acquired resistance to it. Vancomycin kills sensitive bacteria by binding to the growing cell wall and blocking formation of the cross-bridges that give the wall its strength. Resistance arises when a set of three enzymes chemically modifies the wall precursor so that vancomycin no longer binds. Hee-Jeon Hong from Korea, working as a postdoctoral fellow with Mark Buttner (Figure 7.5), found these enzymes to be made only when the organism senses vancomycin, and this adaptive response is regulated by a transcription factor belonging to a class of "two-component" regulators.[7] They control responses to a whole variety of stimuli in virtually all bacteria, and the *S. coelicolor* genome sequence revealed more of them than in any other bacterial genome (about 80, compared with 32 in *E. coli* and 11 in *M. tuberculosis*), another token of the extreme versatility of *Streptomyces* as a soil inhabitant.

These systems each consist of a matched pair of proteins, a sensor and a regulator (Figure 7.6). The sensor lies in the membrane and detects some kind of external influence, in this case vancomycin, which causes a conformational change by binding to the N-terminus of the protein. This transmits a signal via a membrane-spanning region to the C-terminus in the cytoplasm, which in turn changes shape to become an enzyme to phosphorylate one of its own histidine residues (H). Phosphorylation leads the sensor to recognize the regulator component, passing the phosphate group to one of its aspartic acid residues (D) and causing the protein to bind to a DNA sequence upstream of the genes to be switched on. A phosphatase constantly removes the phosphate group from the regulator, so that, when the sensor is no longer adding it back, the regulator leaves its site on the DNA, switching off the response. Such "phosphorelays" are at the heart of many kinds of signaling systems in all forms of life.

Figure 7.5. Hee-Jeon Hong and Mark Buttner at the John Innes Centre, April 2005. (Courtesy of Andrew Davis.)

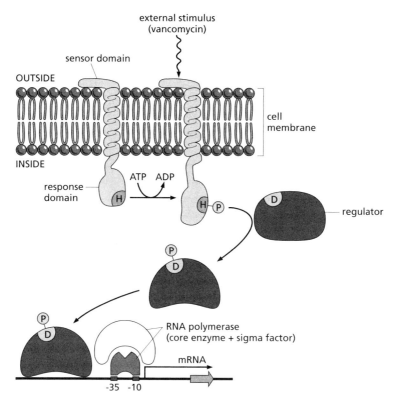

Figure 7.6. The *Streptomyces coelicolor* two-component regulator that detects vancomycin in the medium.

Other stress responses are controlled by alternative sigma factors. *Streptomyces* is better endowed with sigma factors than any other bacterium so far investigated: an amazing 63 of them, compared with 7 in *E. coli* and 17 in *Bacillus subtilis*. Significantly, no fewer than 49 encode members of a group first recognized in *Streptomyces*[8] and now known in almost all bacteria. It is called the ECF family, for *extracytoplasmic function*, because its members are involved in dealing with stimuli or stresses outside the cell, an example being oxidative stress.

All organisms are threatened when atmospheric oxygen produces dangerous chemical groups called free radicals. They cause irreversible damage to enzymes and other cellular components unless their effect is countered. (The beneficial effects of antioxidants such as the pigments of tomatoes and red wine in the human diet highlight the importance of combating oxidative stress.) Oxidative stress changes protein structure. Most proteins contain several molecules of cysteine, one of only two amino acids with a sulfur atom. The sulfur carries a hydrogen atom, forming a sulfhydryl (–SH) group. The correct three-dimensional structure of proteins can depend on pairs of –SH groups joining with each other to form an S–S bridge. These disulfide bonds usually form only in proteins that have been exported from the cell into the hostile outside world, where they provide rigid struts to stabilize the protein. The –SH groups of intracellular proteins usually need to remain free, especially when they are part of the enzyme's active site, but the free radicals that arise during oxidative stress cause the –SH groups of intracellular proteins to form inappropriate disulfide bonds. Cells defend themselves against this damage by expressing a suite of genes that encode protective proteins. Mark Paget, another of Mark Buttner's postdoctoral fellows, in collaboration with the group of Jung-Hye Roe at Seoul National University, Korea,[9] found that one of the *S. coelicolor* ECF sigma factors, sigma-R, ensures that these proteins are produced only when the stress occurs (Figure 7.7). A small amount of sigma-R is present all the time, but it is blocked from joining with the RNA polymerase core by another protein, which binds to sigma-R. This anti-sigma factor has –SH groups that detect oxidative stress by forming an S–S bridge, making the anti-sigma factor change shape and release sigma-R. Sigma-R can now join with the core RNA polymerase to transcribe the set of genes needed to mount the cell's defense, including the gene for sigma-R itself, so the amount of this protein in the cell rapidly increases, amplifying the defense response. Another gene encodes a disulfide reductase, which breaks S–S bridges induced in the cellular proteins by the oxidative stress. The system returns to rest when the disulfide bond on the anti-sigma factor is broken and the protein once more binds to sigma-R, conserving cellular energy by shutting down the defense response.

Switching on Antibiotic Production

Production of antibiotics is highly regulated. They are not made while the young colony is exploiting an abundant food supply, when easily assimilated sources of carbon (e.g., glucose) and nitrogen (e.g., ammonia) and free phosphate repress antibiotic production. Only when the organism begins to sense starvation does antibiotic synthesis start, typically when the aerial hyphae emerge from the substrate mycelium. This behavior is developmentally programmed as an adaptive response, strikingly demonstrated by the finding that most mutants that are defective in forming aerial mycelium, such as the *bld* mutants of *S. coelicolor* (Chapter 6), fail to produce antibiotics.

Why should antibiotics be made only when the colony is ready to sporulate? The answer probably lies in an idea that Keith Chater and Mike Merrick proposed in 1979. On this hypothesis, antibiotic production is primarily not an offensive but a defensive ploy.[10] The aerial mycelium of a *Streptomyces* colony develops at the expense of nutrients supplied by breakdown of the substrate mycelium, as described in Chapter 6. Marauding microbes would threaten these nutrients, and antibiotic production protects them. Keith has taken the hypothesis a stage further and suggested that the sugars, amino acids, and other metabolites produced during the breakdown of the substrate mycelium, which attract many species of motile bacteria, might act like bait. They would encourage competing bacteria to swim into the midst of the *Streptomyces* colony, where they would be killed by the antibiotics and become part of the food source for the developing spores. This sinister scenario might even extend to nematode worms and insects seeking a juicy morsel, which could be knocked out by avermectin or one of several other antibiotics that kill invertebrates and become a

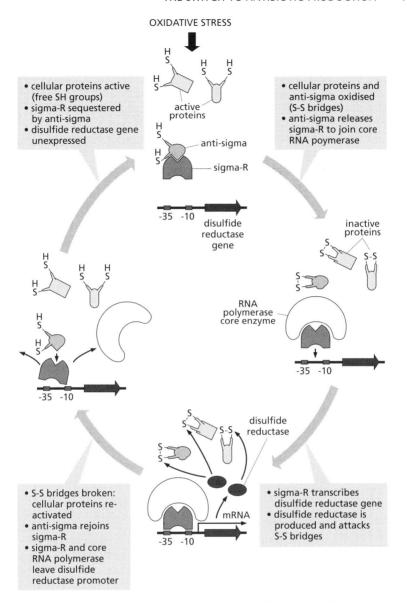

Figure 7.7. The *Streptomyces coelicolor* sigma-R system for dealing with oxidative stress.

meal for the sporulating colony. To be effective over the short distances involved, only small quantities of antibiotic are needed, fitting the difficulty of even detecting antibiotic production in the soil. And because many of the competitors of a *Streptomyces* are other streptomycetes, there has been selection for different strains to make chemically different antibiotics; otherwise their effect would be negated. This neatly accounts for the huge chemical diversity of antibiotics.

Understanding the switch to antibiotic production is more challenging than working out the relatively simple control circuits for utilizing individual nutrients or combating specific stresses. Antibiotic production is triggered and modulated by a complex series of circumstances, a bit like a logic board: "make an antibiotic if x *and* y, but *not* z." All these signals must be interpreted so that the organism can respond in an appropriate way. Regulation therefore involves a cascade of steps that integrates signals from different sources. There is not yet a complete account of any example, but glimpses of parts of several cascades for different antibiotics exist. However, we do know a lot about the antibiotic biosynthetic genes themselves, whose expression is the end-point of the regulatory cascades.

Antibiotic Gene Clusters

Actinorhodin, the compound that gives *S. coelicolor* its beautiful blue color (Color Plate 2B), offered an obvious advantage in the study of the genetics of antibiotic biosynthesis. It was easy to recognize mutants impaired in their ability to make actinorhodin just by a change of color (Color Plate 6). My graduate student, Brian Rudd, isolated 76 such mutants and grouped them into seven classes according to the differently colored compounds made by the various mutants, which were blocked in the biosynthetic pathway at different points. He then tested the mutants in pairs to see whether one would cause another to make the blue antibiotic. When this happened, Brian could deduce the order in the pathway of the various blocks, because a mutant interrupted later in the pathway secretes an intermediate that is converted into the end product of the pathway by an earlier-blocked mutant, but not *vice versa* (Figure 7.8). Mapping of a representative of each class using the natural mating system of *S. coelicolor* showed that the seven genes were close together on the chromosome, suggesting that they might form a contiguous cluster. Brian left John Innes in 1980, one of a number of graduate students and postdoctoral fellows who went on to apply their experience of *Streptomyces* genetics in pharmaceutical companies, in his case Glaxo (now part of the huge GlaxoWellcome).

The story was taken up by Francisco (Paco) Malpartida, who worked with me in Norwich as a postdoctoral fellow from 1982 to 1986 before returning to Madrid to establish his own group (Figure 7.9). Paco was a wizard at the bench and a most thoughtful scientist. He seemed to do a lot of his thinking in the institute car park, where he would pace up and down every few hours in all weathers smoking a particularly strong brand of Spanish cigarettes called *Celtas*, with a Celtic warrior on the packet. He cloned the actinorhodin genes by looking for blue colonies after introduction of random fragments of DNA from the wild type into a blocked mutant on a plasmid cloning vector. Extending Brian's tentative conclusion from his low-resolution mapping, Paco found that all the genes were next to each other on the chromosome. When the DNA was sequenced—a piecemeal effort extending over several years in the era before whole genome sequencing—it was found to encode 22 genes covering 22,000 base pairs[11] (Figure 7.10). Most of the genes encode biosynthetic enzymes for steps in the pathway, including early steps catalyzed by a polyketide synthase of the kind discussed in detail in Chapter 8, as well as "tailor-

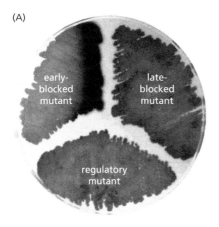

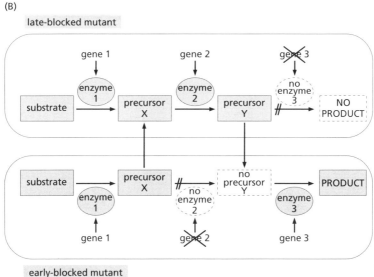

Figure 7.8. Cosynthesis of actinorhodin in a pairwise combination of *Streptomyces coelicolor* mutants blocked at different points in the biosynthetic pathway to the antibiotic. (A) Dark-colored actinorhodin is made by an early-blocked mutant when an actinorhodin precursor diffuses to it from a late-blocked mutant; no cosynthesis occurs with the regulatory mutant, in which the entire biosynthetic pathway is switched off. (B) Explanation of cosynthesis.

ing" steps later in the pathway. There is a gene conferring self-resistance near the middle of the cluster; it encodes a protein that resembles the tetracycline export pump.

Although the cluster of genes for actinorhodin biosynthesis is on the *S. coelicolor* chromosome, those for another antibiotic, methylenomycin, are carried on the SCP1 plasmid.[12] When Ralph Kirby and Fred Wright, then two graduate students in my laboratory, discovered this in 1975, it caused a stir because, if this had turned out to be a general phenomenon, it might have provided a rapid strategy for improving the

Figure 7.9. Francisco (Paco) Malpartida at the Institute of Biotechnology, Autonomous University of Madrid, March 1999. (Courtesy of Universidad Autónoma de Madrid.)

productivity of industrial cultures. After a strain had been laboriously mutated to make more of a particular antibiotic, a plasmid carrying the genes for a different antibiotic might perhaps have been transferred by mating into the improved strain to cause it to overproduce the compound. However, although Haruyasu Kinashi and his group in Hiroshima recently described another example of plasmid-determined antibiotic biosynthesis—a plasmid of *Streptomyces rochei* carries no fewer than three clusters of antibiotic biosynthetic genes[13]—the many other clusters studied over the last 25 years have turned out to be chromosomal, including two more antibiotics of *S. coelicolor*. One of these is red and quite insoluble in water, so it does not diffuse far into the medium. It has the rather horrendous name of undecylprodigiosin and is usually referred to as "the red antibiotic." The third antibiotic is colorless and is known as the calcium-dependent antibiotic (CDA) because it kills bacteria only if calcium is present in the medium.

Almost all antibiotic gene clusters resemble the actinorhodin cluster in that they contain one or more genes for self-resistance. Obviously the organisms need such genes, but why should they be embedded with the biosynthetic genes, rather than being somewhere else in the genome? The answer probably reflects the fact that the capacity for antibiotic biosynthesis has been passed around between different actinomycetes over evolutionary time, in the process of horizontal gene transfer described in Chapter 5. For this to lead to viable progeny, resistance would need to be inherited along with antibiotic production; otherwise, recipients of the biosynthetic genes would not survive.

Most antibiotic gene clusters also contain regulator genes whose products activate transcription of all the biosynthetic genes in the cluster by binding to DNA sequences upstream of them, as Paco's group showed for the actinorhodin activator.[11] That activator belongs to a family of transcription factors later found in many

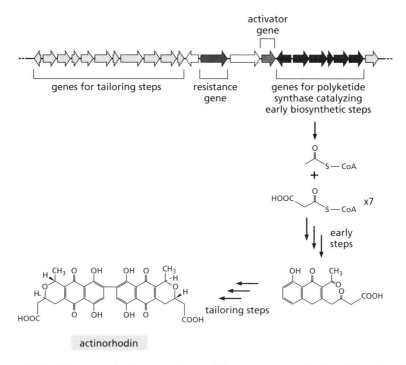

Figure 7.10. The actinorhodin gene cluster of *Streptomyces coelicolor*. (The roles of the polyketide synthase genes are described in Chapter 8.)

actinomycete gene clusters, including those for the red antibiotic and CDA in *S. coelicolor,* each being specific for expression of the genes in its own cluster. Andreas Wietzorrek, a postdoctoral fellow in Mervyn Bibb's group in Norwich, discovered the basis for this specificity when he found that the sequences to which the activators bind differ slightly in the various clusters.[14]

Some of the genes in the actinorhodin cluster are transcribed individually into messenger (mRNA) molecules, and others are transcribed together in functional units called operons. The definition of the operon arose from the genetic work of Jacob and Monod in Paris (Chapter 6), who found that the *E. coli* transporter for lactose and the enzyme needed to metabolize it were always expressed together in a constant ratio. Soon it was established that bacteria often express two or more genes from the same mRNA molecule, regulated by the same promoter and operator. The operon is one of the most important concepts in bacterial genetics. The 22 genes in the actinorhodin cluster are organized into half a dozen operons, and this is typical for *Streptomyces* antibiotic gene clusters.

Many such clusters have now been analyzed in great detail, an important one being the cluster for streptomycin in Waksman's *Streptomyces griseus* strain. As shown by Wolfgang Piepersberg and his group at the University of Wuppertal, Germany, this cluster consists of more than 30 genes, reflecting the complex structure of the antibiotic.[15] The cluster contains a resistance gene that protects the cell by keeping streptomycin in its inactive phosphorylated form until it leaves the mycelium, and an activator belonging to a different protein family from the one for actinorhodin.

What are the steps in the cascade that ends in transcription of the biosynthetic and resistance genes when a pathway-specific activator is expressed? We do not know the full story. However, an important finding is that the activator of the actinorhodin cluster contains a TTA codon, making it a potential target for developmental regulation by the *bldA* gene described in Chapter 6 and suggesting a reason why the *bldA* mutant does not make actinorhodin. Sure enough, when Miguel Fernández-Moreno in Paco Malpartida's laboratory in Madrid changed the TTA to a commonly used leucine codon, a *bldA* mutant produced actinorhodin.[16] Regulatory genes in the clusters for the red antibiotic and for methylenomycin also contain TTA codons, underlining the importance of the *bldA*-mediated switch.

One of the challenges faced by *Streptomyces* growing in the soil is to coordinate antibiotic production in different regions of the colony. It would be futile for just a few hyphae to start making an antibiotic in isolation. It has to be a group activity, but the colony consists of hyphae of different ages. Those at the periphery are young and still foraging for food, although it is probably an increasingly fruitless search, because the colony is likely to have exhausted its immediate surroundings of readily assimilated nutrients. Inside the colony, the older hyphae are ready to support aerial mycelium development and sporulation and therefore need to produce antibiotics to protect their resources from competitors. They coordinate the colony by producing hormones that signal to the younger cells that it is time for them too to start to produce antibiotic and switch to sporulation.

Bacterial Hormones

Alexander Stepanovich Khokhlov (1916–1997) discovered the first example of a bacterial hormone, in Waksman's *S. griseus* strain. After serving as chief of Chemistry at the All-Union Research Institute for Antibiotics in Moscow, he headed a new Institute for the Chemistry of Natural Products, where he and his team discovered what they called the A-factor (for Autoregulator) in the early 1960s. Certain *S. griseus* mutants had lost the ability to make antibiotic and to sporulate, like the *S. coelicolor bld* mutants studied later, and the two abilities were restored when a culture filtrate of the wild type was added to them. The compound responsible turned out to be a novel type of molecule called a gamma-butyrolactone, with a five-membered ring containing an oxygen atom and four carbons, as well as an oily tail (Figure 7.11A). Chemical characterization of A-factor was a challenge because only minute quantities were produced, so the researchers had to purify it from very large volumes of culture medium.

Two Japanese groups, headed by Teruhiko Beppu in Tokyo and Yasuhiro Yamada in Osaka, took up the study of A-factor and related compounds from the early 1980s onward. Beppu worked on A-factor in *S. griseus*, and he and his successor at the University of Tokyo, Sueharu Horinouchi (Figure 7.12), went on to describe a detailed molecular cascade by which it activates streptomycin production. Yamada found several other members of the same chemical family of microbial hormones in

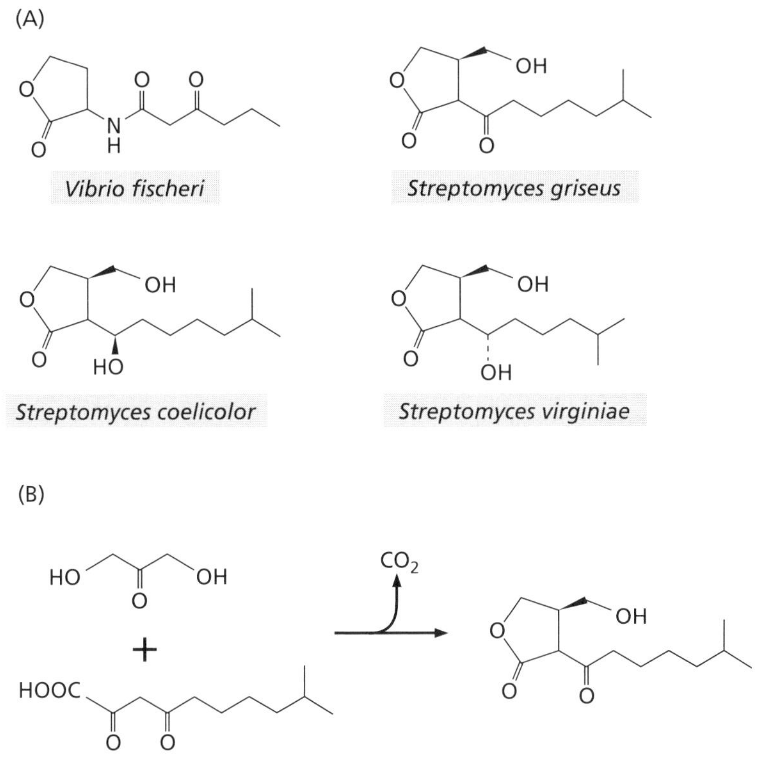

Figure 7.11. (A) Gamma-butyrolactones from three different *Streptomyces* species, including the classic A-factor made by *Streptomyces griseus*, compared with the homoserine lactone of *Vibrio fischeri*. (B) Proposed biosynthesis of the *S. griseus* A-factor from glycerol and alpha-ketoglutarate.

Streptomyces virginiae, where they triggered virginiamycin production. More recently, especially because of the availability of its genome sequence, attention has turned also to *S. coelicolor*, which produces yet other members of the same class of compounds. The gamma-butyrolactones are widespread hormones in *Streptomyces*.[17]

Before carrying on with the story, let us look in the deep seas, where various kinds of fish and molluscs have developed special organs, often just below the eyes, that emit light (Color Plate 7). These organs help in catching prey by illuminating them or attracting them with a lure, in advertising for a mate, or in providing camouflage. The interesting thing is that the animal itself does not generate the light. The light organs are home to gram-negative bacteria of the genus *Vibrio*, and it is they that glow in the dark. They do this using an enzyme called luciferase, which converts energy derived from a chemical reaction driven by ATP to bluish-green light. Light production does not make obvious sense for bacteria living free in seawater. It happens when a dense population of *Vibrio* occupies a limited space, and a cell density–dependent mechanism has evolved to regulate the relevant genes, as first described by scientists at the Scripps Institution of Oceanography at La Jolla, California, in 1977.[18]

Figure 7.12. Teruhiko Beppu at a Max Tishler Memorial Symposium, November 10, 1999 (courtesy of the Kitasato Institute, Tokyo), and in conversation with Sueharu Horinouchi at Oxburgh Hall, Norfolk during a Japan–U.K. Joint Meeting on Molecular Genetics of *Streptomyces*, October 26, 1997. (Courtesy of Tobias Kieser, John Innes Centre).

This regulation works through hormones called quorum-sensing molecules, homoserine lactones, very similar to the gamma-butyrolactones of *Streptomyces* but with a nitrogen at a position where the gamma-butyrolactones have carbon (Figure 7.11A). The *Vibrio* probably produces the hormone all the time, but in seawater it diffuses away and is lost in the ocean. When a population of bacteria is confined in the light organ, the hormone concentration builds up and triggers the light response. It binds to a transcription factor, enabling it to activate a set of genes encoding luciferase, as well as enzymes for making the chemical substrate that luciferase needs to generate light. There is a built-in amplification device, because three genes that control steps in the biosynthesis of the homoserine lactone are also switched on, and the low basal level of the hormone suddenly rises. The regulatory system is crucial to the economy of the bacteria, because luciferase can account for at least 5% of total cellular protein, and light can represent 10% of total energy production when the genes are fully expressed, so it would be extremely wasteful for this to happen outside the light organ. But inside it, light production benefits the host, which in turn provides the bacteria with a home and a food supply.

Different homoserine lactones were later found to play roles in gene regulation in various other gram-negative bacteria, for example in pathogenesis by *Pseudomonas aeruginosa*. Again, the quorum-sensing response is adaptive, because it is useless for a bacterium to mount an attack on the host's defenses unless it can muster enough forces to breach them.

The regulatory circuits controlling antibiotic production in *Streptomyces* resemble the one for light emission by the *Vibrio* but are more complex: several transcription factors act as repressors and activators in a cascade, as explained in Figure 7.13 for *S. griseus*. An enzyme makes the A-factor (Figure 7.11B) in the older vegetative

THE SWITCH TO ANTIBIOTIC PRODUCTION   163

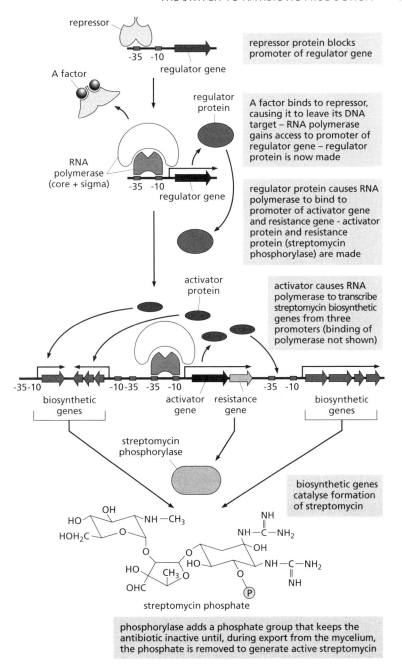

Figure 7.13. The A-factor cascade regulating streptomycin biosynthesis in *Streptomyces griseus*.

Figure 7.14. Eriko Takano in Norwich, November 2002. (Courtesy of Tobias Kieser, John Innes Centre.)

mycelium that is ready to produce streptomycin and needs to signal to the younger hyphae to do the same. The hormone binds to a repressor, causing it to lose its affinity for a target sequence upstream of the next regulator gene. This can now be expressed, leading in turn to transcription of another gene in the cascade, which produces an activator that switches on transcription from the various promoters in the streptomycin cluster to express the biosynthetic enzymes, as well as the resistance protein for self-defense, and streptomycin is made.

Different *Streptomyces* strains produce different members of the gamma-butyrolactone family with different specificities. The chemical variations are in the fatty acid side chains (Figure 7.11A). The *S. griseus* A-factor hardly affects *S. coelicolor*, and a factor identified in *S. coelicolor* by Eriko Takano in Norwich (Figure 7.14), in collaboration with the Osaka group, has no effect in *S. griseus*. Because many of the competitors for a *Streptomyces* in the soil are other *Streptomyces* species, it makes sense for different specificities to have evolved. In *S. coelicolor,* the antibiotic regulatory cascade is even more complex than in *S. griseus*,[19] perhaps because this organism makes four antibiotics, in contrast to the single antibiotic made by the streptomycin-producing *S. griseus* strain.

As we have seen in this chapter, antibiotic production by *Streptomyces* is a complex reaction to the stresses the organisms face in their natural habitats. Its ongoing study will continue to illuminate the activities and life styles of soil microorganisms, but meanwhile it has also affected biotechnology. It has offered a route to making novel antibiotics by doing chemistry through genetics, supplementing the more conventional kind practiced by chemists in laboratories. The next chapter describes this new approach to finding valuable drugs.

8

Unnatural Natural Products

After a couple of lean decades in which no useful natural antibiotics were discovered (Figure 2.1), most drug companies concluded that screening of microorganisms was no longer a worthwhile strategy. Two alternatives offered to fill the gap. One was to return to synthetic chemistry. Before the antibiotic era, chemists had dominated the pharmaceutical industry. The arsenic compound Salvarsan was an unpleasant treatment for syphilis in the 1920s, but the synthetic sulphonamides that immediately preceded the antibiotic era were very good antibacterials. Although largely eclipsed by penicillin, they still have their uses. The early 1990s saw a renewed assertiveness by medicinal chemists, emboldened by the invention of a new field of synthesis, called combinatorial chemistry, in which robots make thousands of compounds in miniaturized reactions, ringing the changes on a set of building units joined in different ways. It is a bit like letting a horde of preschool children loose on a mound of Lego blocks to see how many different structures they can build in a morning's session. Vast numbers of new compounds can come out of a "combichem" laboratory, and a few promising leads are being followed up, but most synthetic compounds will not interact with microbes or cell receptors in interesting ways, so the hit rate is very low.

The alternative is to use genetic engineering to make "unnatural natural products"—unnatural because they have not been found in nature, and natural because they are made by microbes rather than laboratory chemists. Antibiotics are complex molecules, usually with one or more "asymmetric centers" (carbon atoms carrying four different chemical groups), which therefore exist in two mirror-image forms (stereoisomers), only one of them biologically active. Synthetic chemists struggle to make pure stereoisomers, but microbes have evolved over eons to build them precisely; stereospecificity is a hallmark of enzyme-catalyzed reactions. Making unnatural

natural products became possible as soon as genetic manipulation in *Streptomyces* was up to the challenge of isolating and analyzing gene sets for antibiotic biosynthesis. This chapter describes how this new field of biotechnology arose.

Hybrid Antibiotics

One May morning in 1984, Helen Kieser and I examined some cultures she had inoculated a few days earlier. Some were beginning to develop the blue color characteristic of *Streptomyces coelicolor* making actinorhodin, and others were brown, but one was a shade of purple we had never seen before (Color Plate 8). We looked at each other excitedly. It was the sort of result we had dreamed of when we set out to try to use the newfound power of genetic engineering to generate a novel antibiotic. Why were we so excited about this color change?

By this time, Paco Malpartida had cloned the actinorhodin biosynthetic genes. Only later did the DNA sequence reveal the exact number and arrangement of the genes (Figure 7.10), but meanwhile Paco proved that the set was complete by transferring the clone to *Streptomyces parvulus,* which makes no compounds similar to actinorhodin. It made the antibiotic. Now we needed some *Streptomyces* species producing other members of the chemical family to which actinorhodin belongs—the benzoisochromanequinones, or BIQs for short—to see if the actinorhodin genes would cause them to make novel compounds. Would intermediates along the pathway for one antibiotic be hijacked by the enzymes of another pathway and converted into a hybrid end product?

In February 1984, I wrote to my friend Satoshi Ōmura at the Kitasato Institute, suggesting he send some of his BIQ-producing strains so that we could introduce the actinorhodin genes into them and send them back to Tokyo for chemical analysis. By mid-April several cultures had arrived, including one that made the brown-pigmented antibiotic medermycin. This molecule differs from actinorhodin in three ways (Figure 8.1): it has a sugar attached to the carbon ring system instead of consisting of two copies of the set of rings joined back to back; it lacks a hydroxyl (–OH) group at one position on the ring system; and it contains a fourth ring instead of the free carboxyl (–COOH) group of actinorhodin. Helen made protoplasts of the medermycin-producing strain and introduced into them a set of Paco's clones carrying overlapping segments of the actinorhodin gene cluster. It was the result of this experiment that made us so excited. We thought the new purple pigment must be a hybrid compound with some features of the brown medermycin molecule and others of the blue actinorhodin, but proof was needed. On May 5, 1984, I sent the purple culture to Satoshi, and by early August back came a letter telling us they had proved the hypothesis: the purple compound was indeed a chemical hybrid resembling medermycin but with the extra -OH group of actinorhodin (Figure 8.1). We acknowledged this by calling it mederrhodin. Soon we were drafting a letter to *Nature* magazine.

Meanwhile, I had contacted another friend, the well-known natural product chemist Heinz Floss, at Ohio State University, Columbus. I suggested that we introduce the actinorhodin genes into a strain of *Streptomyces violaceoruber* that makes a BIQ called dihydrogranaticin, which has a beautiful, psychedelic purple color. Heinz had proved many of the steps in its biosynthesis, including how a special sugar is made and fused

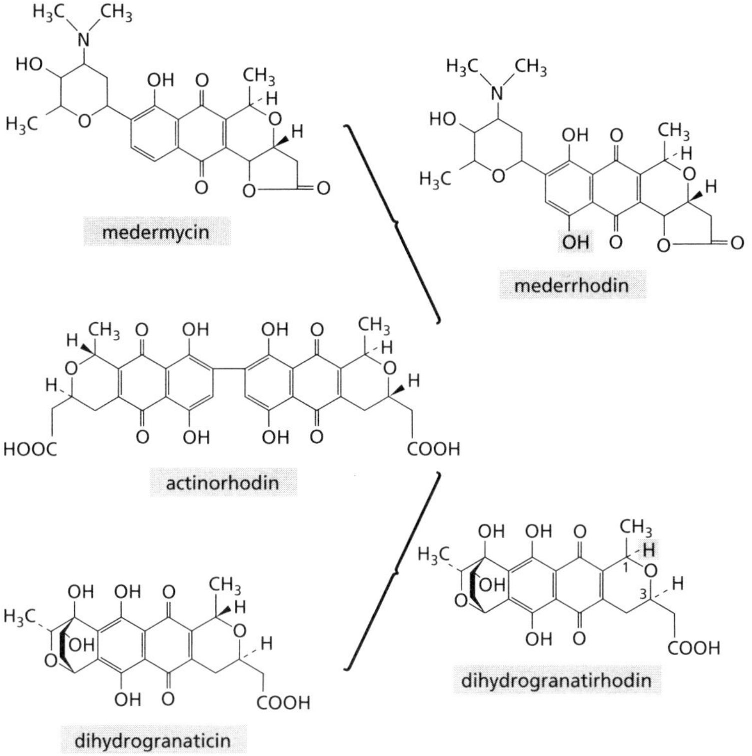

Figure 8.1. Structures of medermycin, actinorhodin, and dihydrogranaticin, together with those of the hybrid antibiotics mederrhodin and dihydrogranatirhodin, with changes highlighted.

to the carbon ring system in an unusual way (Figure 8.1). It was hard to persuade this strain to take up DNA, but Helen managed to introduce a few of Paco's actinorhodin clones into it, including one carrying the complete gene cluster. At the end of June, I sent this strain to Heinz for analysis. There was no new color to indicate that a hybrid compound might be made, and the first news from Columbus was unpromising, but hard work by the Floss group was rewarded. On November 7, I had an excited phone call from Heinz. They had detected a hybrid compound differing from dihydrogranaticin in the stereochemistry in the right-hand ring. The BIQs have two asymmetric centers, with opposite stereochemistry in actinorhodin and dihydrogranaticin. In the new compound, one center had the stereochemistry of dihydrogranaticin and the other that of actinorhodin (Figure 8.1). We gave it the rather clumsy name of dihydrogranatirhodin.

This result came in time for the Floss group to be included in the *Nature* article, which came out on April 18, 1985.[1] I met Heinz and Satoshi again soon afterward, the first time we had talked about our three-way collaboration face to face (Figure 8.2). The article sparked a lot of interest: the UK's Central Office of Information promoted the story and sent us press cuttings about our work from faraway places, such as *Arab*

Figure 8.2. Heinz Floss, David Hopwood, and Satoshi Ōmura at a conference on *Regulation of Secondary Metabolite Formation* held at Gracht Castle, Germany, May 12–16, 1985. (Courtesy of Horst Kleinkauf, Technical University, Berlin.)

News in Jeddah, Saudi Arabia, *Trinidad Guardian* of Port of Spain, and *El Pueblo* of Montevideo, Uruguay, as well as more familiar papers from nearer home. I had talked about mederrhodin at the British Association for the Advancement of Science, which happened to hold its annual meeting in Norwich in September 1984, and there was a nice writeup in the *Economist* magazine by their science correspondent Matt Ridley, who has gone on to be one of the best science writers in the United Kingdom (Figure 8.3). He suggested the pharmaceutical industry should take note of our demonstration that unnatural antibiotics could be made by genetic engineering, as a new approach to fighting infections and cancers.

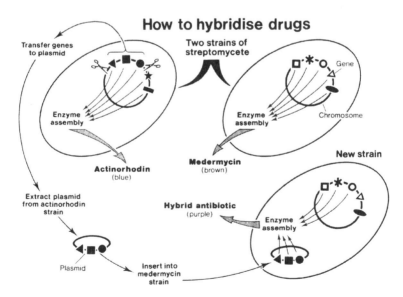

Figure 8.3. Cartoon from *The Economist* illustrating the engineering of the first hybrid antibiotic. (Copyright The Economist Newspaper Ltd., London, September 22, 1984.)

In our experiments, we had simply exploited the colors of the compounds as convenient flags for the underlying genetics and chemistry to see whether hybrid antibiotics could be made, not to yield useful drugs. And mederrhodin and dihydrogranatirhodin differ only a little from their parent molecules, although these changes would be hard to achieve faithfully by chemical synthesis. If useful antibiotics were to be made, more fundamental changes in the structures of natural products would need to be engineered. It turned out the BIQs had been a good choice for our experiments, because they belong to a superfamily of natural products, the polyketides, with a huge range of interesting biological activities. Almost half the compounds in Table 2.1 are polyketides, including most of the important antibacterial, antifungal, anticancer, and immunosuppressive agents. The superfamily also includes plant pigments, fungal toxins, and bizarre compounds with doubtful biological roles from marine microorganisms, many living inside sponges in warm waters such as the Caribbean Sea; some of these are promising anticancer agents. It is amazing that all of these compounds, which differ so widely in their chemical structure and hence their biological properties, are made by ringing the changes on the way that organisms make fatty acids, the simple oily hydrocarbon chains that build cell membranes.

Polyketide Biosynthesis

Fatty acids and polyketides are made by complex "synthase" enzymes, which build them by linking simple precursors into carbon chains ranging from fewer than 10 to several hundred atoms long. The building units come from organic acids, of which acetic acid, the acid in vinegar, is the simplest. As often happens in biochemistry, they have to be attached to a chemical carrier to make them reactive, and in the case of fatty acid and polyketide biosynthesis this is coenzyme A, or CoA. Thus we speak of acetyl-CoA as the building unit derived from acetic acid, which is attached to a sulfur atom on the coenzyme.

The reaction linking two units to form a new carbon–carbon bond is called a condensation, because it removes two hydrogen atoms and one atom of oxygen in the form of water from the molecules being joined. To lengthen the carbon chain by two atoms, the extender unit must have at least three carbons, because one is lost as carbon dioxide to drive the reaction, so the simplest extender unit is the three-carbon malonic acid, in the form of malonyl-CoA. To participate in condensation, the malonyl unit is first transferred from CoA to a component of the polyketide synthase (PKS) or fatty acid synthase (FAS) called an acyl carrier protein (ACP) and tethered to it via a sulfur atom. Meanwhile, the starter unit is transferred from its CoA derivative to the active site of the enzyme and attached to the –SH group of a cysteine. The transfer and condensation reactions are shown in Figure 8.4.

Next comes a second round, in which the four-carbon product of the first condensation is transferred back to the active site of the enzyme and condensed with another malonyl unit attached to the ACP, extending the chain to six carbon atoms. By further rounds of condensation, the chain grows to an appropriate length and is removed from the ACP (Figure 8.5).

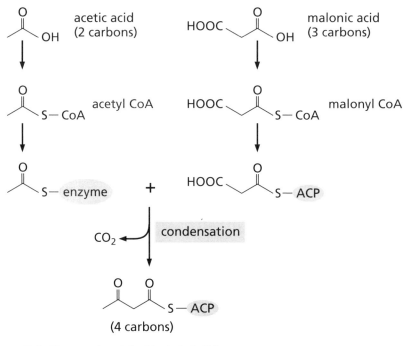

Figure 8.4. First step in polyketide chain building.

Notice that the end product has an oxygen atom attached by a double bond as a keto-group to every alternate carbon atom in the chain: this is why these compounds are called polyketides.

If this were the whole story, polyketides would vary only in the length of the carbon chain, but there are other variables. The keto-groups do not all remain intact but may be modified by a succession of three enzyme-catalyzed steps. The first is reduction (addition of hydrogen) by an enzyme called a ketoreductase, which converts the keto-group to a hydroxyl. Next, a dehydratase removes the hydroxyl group and a hydrogen from the adjacent carbon as a water molecule to produce a double bond in the carbon chain. Then an enoyl reductase adds hydrogen to each of the carbon atoms joined by the double bond, converting it back to a single bond (Figure 8.6).

The sequence of three steps that modifies the keto-groups can stop at any point, so we find a keto-group at some positions in the chain, a hydroxyl group elsewhere, a double bond between some pairs of carbon atoms, with no side group at other positions (Figure 8.7).

There are other sources of variation too. A range of organic acids other than acetyl-CoA can start the chain. Some of them are shown in Figure 8.8.

The extender units also can vary. Instead of malonyl-CoA, with three carbon atoms, the extender can be methylmalonyl-CoA, with four carbons. Each step of chain building always adds just two carbon atoms to the length of the chain, and the extra

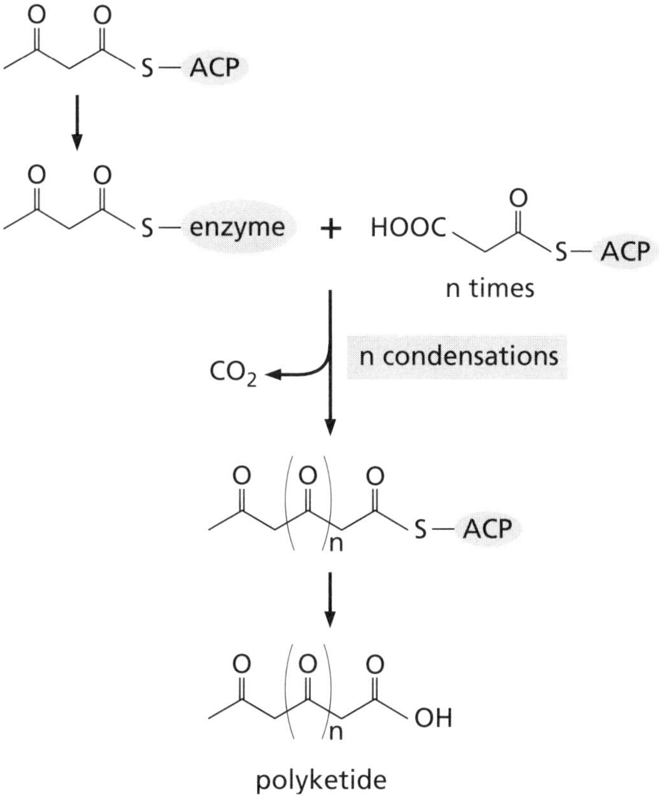

Figure 8.5. Subsequent steps in polyketide chain building.

carbon forms a side-branch as a methyl group. Other extender units result in more complex branches (Figure 8.9).

As a further source of variation, wherever there is a hydroxyl group produced by the ketoreductase, or a carbon branch resulting from the incorporation of an extender unit other than malonyl-CoA, it can have either of two possible stereochemical arrangements. All these variables—choice of starter unit, choice of extender unit, chain length, fate of each keto-group, and stereochemistry of the hydroxyl and carbon branches—affect the shape and chemical properties of the molecules and therefore their biological activity. And these factors are almost all independent, so the number of combinations and permutations is astronomic. Other reactions may modify the carbon chain after it is built, for example by folding it into rings, and sugar molecules or other chemical groups may be attached by tailoring enzymes. All this variation explains how polyketides play such a huge variety of biological roles.

Different components of the PKS or FAS perform the six reactions that it carries out. There are two acyl transferases that recognize the correct starter and extender units and attach them to their sites on the enzyme; a ketosynthase catalyzes conden-

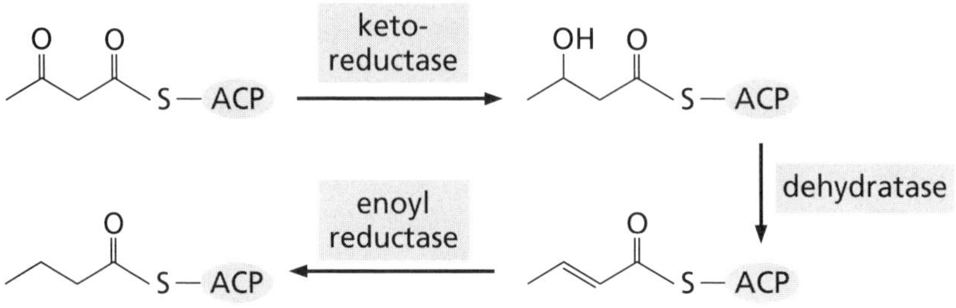

Figure 8.6. The reductive cycle on ACP.

sation; and the ketoreductase, dehydratase, and enoyl reductase act on the keto-groups. The synthases also include the small ACP, to which the carbon chain is tethered via a sulfur atom on a flexible arm that is long enough to carry the carbon chain to the various active sites that work on it. Thus the full repertoire of the PKS or FAS consists of seven functions.

By the time we started working on the genetics of polyketides, biochemists had discovered that these functions are arranged differently in a bacterial FAS compared with one from a eukaryote.[2] In *Escherichia coli*, separate genes encode individual proteins that come together as an aggregate. In contrast, the FAS of vertebrates consists of one large protein encoded by a single gene, with the different functions catalyzed by "domains" along the protein. Separate genes were doubtless joined together at some stage along the evolutionary path from the progenitors of present-day bacteria to those of eukaryotes to generate the vertebrate FAS. The PKS that makes a fungal metabolite called 6–methyl salicylic acid had been found to resemble a vertebrate FAS, with the various active sites on one protein, and even arranged in the same order, providing striking evidence that FAS and PKS enzymes have a common evolutionary origin.

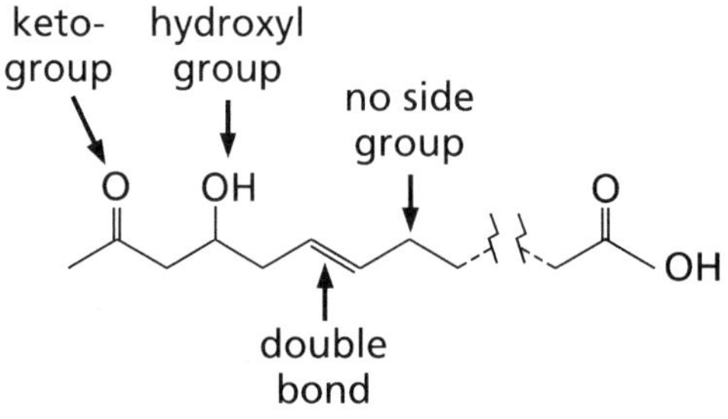

Figure 8.7. A reduced polyketide.

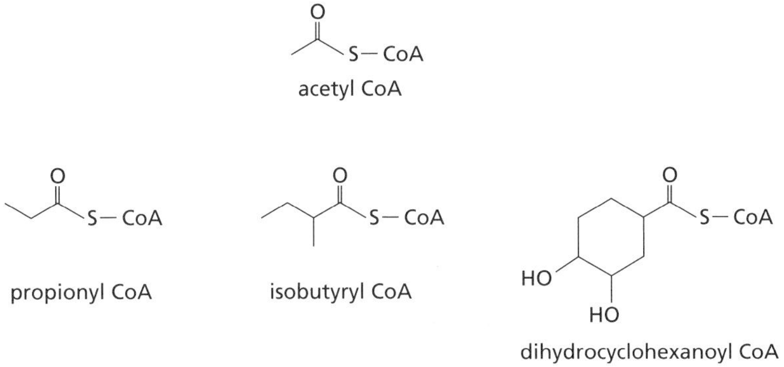

acetyl CoA

propionyl CoA   isobutyryl CoA   dihydrocyclohexanoyl CoA

Figure 8.8. Starter units for polyketide biosynthesis.

Fatty acids are simpler than most polyketides because they are made using a limited set of the possible variables, the main one being chain length. Most fatty acids start with acetyl-CoA, and use malonyl-CoA for every extension, so there are no carbon branches on the chain. Almost all the keto-groups are fully reduced, so there are usually no keto- or hydroxyl side groups, and there are either no double bonds in the chain (in the saturated fatty acids) or just one or a few in the famous unsaturated fatty acids that make up a healthy diet. Furthermore, fatty acids do not fold into complex carbon ring systems after the chain is built. In contrast, a PKS is like a computer, "programmed" to make a series of choices in building a particular polyketide. By the end of the 1980s, the time was ripe to exploit the new powers of cloning and recombining *Streptomyces* genes for polyketide biosynthesis to try to understand the programming software and then change it experimentally to make designer polyketides. We could not have embarked on this endeavor without a series of timely visits from U.S. researchers.

David Sherman (Figure 8.10) came to Norwich in 1987 from the biotech company Biogen in Cambridge, Massachusetts, where he had converted from chemistry to mammalian cell genetics. He and his wife arrived at London's Heathrow Airport

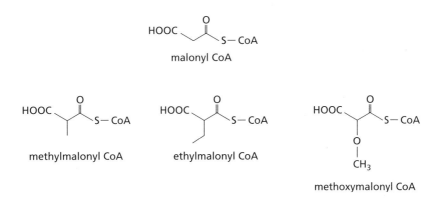

Figure 8.9. Extender units for polyketide biosynthesis.

Figure 8.10. David Sherman with a culture showing an early example of complementation between dihydrogranaticin and actinorhodin PKS subunits at the John Innes Centre, 1991. (Courtesy of Tobias Kieser.)

early one August morning, planning to stay for 3 years, but initially they were denied entry to the United Kingdom because the immigration officer thought David would be working illegally. I was in the United States at a meeting, so the administrators at the John Innes Centre had to respond to frantic telephone calls as best they could with the information they had. Immigration was not convinced, but the Shermans were allowed to stay in the country for a few days as a concession. Before their time was up, I drove them back to Heathrow, where a lawyer engaged by the John Innes Centre met us. Eventually the immigration officer was convinced that the way we were planning to finance David's stay was completely legal, and the episode ended in handshakes all round.

David sequenced the genes for the dihydrogranaticin PKS, working with DNA that Paco Malpartida had isolated using the corresponding region of the actinorhodin gene cluster as a hybridization probe, a method that became widespread for finding PKS genes in many different actinomycetes. In parallel, Mervyn Bibb took sabbatical leave from Norwich at the Cetus company in Emeryville, California, and sequenced the corresponding DNA from the producer of another polyketide, tetracenomycin, made by *Streptomyces glaucescens*, in a collaboration with Richard (Dick) Hutchinson, whom he had come to know while Dick was on sabbatical with me in Norwich in the second half of 1983. Dick (Figure 8.11) was a major player in polyketide chemistry and bio-

Figure 8.11. Richard (Dick) Hutchinson in the Hortobagy National Park, Hungary during the Sixth International Symposium on Biology of Actinomycetes (ISBA'85), August 1985. (Courtesy of Mervyn Bibb, John Innes Centre.)

synthesis at the University of Wisconsin, Madison. My PhD student, Stephanie Hallam, had already sequenced part of the corresponding region of the *act* gene cluster of *S. coelicolor*, and Paco's laboratory in Madrid sequenced the rest. All three PKS sequences—for dihydrogranaticin, tetracenomycin, and actinorhodin—revealed a set of genes encoding separate proteins, some clearly resembling components of the *E. coli* FAS.[3] We could easily identify genes corresponding to the ketosynthase and the ACP because of their strong similarity to the corresponding *E. coli* proteins. This immediately told us that the PKSs for the *Streptomyces* polyketide antibiotics are organized in the same way as the *E. coli* FAS. But how was the programming controlled? Genetic mix-and-match experiments provided vital clues.

Combinatorial Biosynthesis

In dihydrogranaticin and actinorhodin, one of the keto-groups is reduced to a hydroxyl, whereas tetracenomycin is unusual in having no ketoreduction. It was therefore encouraging that the PKS clusters for the first two compounds included a ketoreductase gene and that the tetracenomycin cluster lacked one, but this was only a small step toward understanding the programming rules. We decided to make hybrid gene clusters and try to deduce the rules from the structures of compounds they made: Which component of the enzyme determined chain length, which influenced the extent and position of ketoreduction, and how was the starter unit chosen? But would the hybrid enzymes even work?

As soon as David Sherman had sequenced the dihydrogranaticin PKS gene cluster, he and Maureen Bibb (Mervyn's wife) embarked on a very informative experiment, which David continued after leaving Norwich for a faculty position at the

University of Minnesota in Minneapolis. When they introduced the various dihydrogranaticin genes into mutants defective in the corresponding components of the actinorhodin PKS, the hybrids made pigmented compounds. This proved that the two gene sets encoded proteins that were similar enough to form a functional hybrid enzyme in which the "good" protein encoded by the dihydrogranaticin gene made up for the defective version of the actinorhodin protein; evidently they had not diverged too much from a common ancestor to recognize each other and form an active enzyme.[4] Bill Strohl's group at Ohio State University, in a collaboration with Heinz Floss, now at the University of Washington in Seattle, had just described another example of such "complementation," between an actinorhodin PKS gene and the set of genes for an anticancer compound called aclacinomycin.[5] These results were very encouraging and laid a firm foundation for producing many hybrid PKSs by mixing and matching the component proteins.

My next postdoctoral fellow from America was Chaitan Khosla. He had studied chemical engineering at the Indian Institute of Technology in Bombay and gone to Caltech in Pasadena to join the group of Jay Bailey, an expert in metabolic engineering. For his PhD, Chaitan studied an interesting gene from a bacterium called *Vitreoscilla,* which lives in mud at the bottom of freshwater pools. It encodes a protein resembling the hemoglobin in our blood that enables the bacterium to scavenge oxygen when the water becomes oxygen starved. The idea was to put the gene into organisms to be grown in industrial fermenters, in the hope that the hemoglobin would reduce the need to pump air into the tanks and thereby save a lot of energy. Jay set up a small company called Exogene to exploit the gene, which was found to work in laboratory-scale cultures but was not in the end commercialized.

I first met Chaitan in September 1988 at a conference of the American Chemical Society in Los Angeles, where I had been invited to give a lecture in memory of David Perlman, one of the best-known figures linking pharmacy with the fermentation industries after the Golden Age. In July of that year, I had spoken at the European Congress of Biotechnology in Paris about our hopes for generating novel polyketides. The proceedings had been recorded, and Jay had given the tape of my lecture to Chaitan to see if he would be interested in moving into the field. Talking to Chaitan over breakfast in my hotel in Los Angeles—he had cold-called me from the lobby to follow up an application to join my laboratory which I had not yet received—I was so impressed with his work and his ideas that I offered him a job without waiting for a *curriculum vitae* or references. It was one of the best decisions of my career.

Chaitan came to Norwich in January 1989 and was there for almost 2 years before moving to a faculty position at Stanford University. His stay therefore overlapped with the second half of David Sherman's. They were joined by two other American chemists who came on sabbatical to learn *Streptomyces* genetics and use its power to solve the biosynthetic problems they were grappling with.

Stephen Gould, on sabbatical from Oregon State University at Corvallis, spent a year in the laboratory beginning spring 1989. He was studying polyketide antibiotics with unusual, angled carbon ring systems, called kinamycins, which are made by *Streptomyces murayamaensis*, and wanted to isolate the biosynthetic genes using Paco's actinorhodin probes. David Cane, from Brown University in Providence, Rhode Island, was another leader in polyketide chemistry (Figure 8.12). He was on

Figure 8.12. David Cane at a meeting on "Secondary Metabolites: Their Function and Evolution" at the Ciba Foundation, London, February 19, 1992. From left to right: Peter Leadlay, Dudley Williams, David Cane, Arnold Demain, Iain Hunter. (Copyright David Ridge Photography.)

sabbatical in Cambridge, United Kingdom, during the academic year 1989–1990, and from January to June would drive over to Norwich several times a week to try to isolate genes for making another class of natural products on which he was a world expert, the terpenoids. These are especially abundant in fungi and plants, where they have important roles as carotenes and steroids, but a few are actinomycete metabolites, and he tried to isolate the gene for a key enzyme in one of these pathways.

David Cane did not succeed in cloning the gene he wanted while he was in Norwich—that came later after his return to Providence—but he found another interesting gene instead. It did not faze him too much: "You may not get the clone you love but you'll love the clone you get." He had a wonderful gift for repartee and would keep us in stitches with an endless stream of apt one-liners and jokes. He had had an English postdoctoral fellow called Chris Abell, and one of David's satisfactions was to have been first author on a paper with Chris. When David introduced Chris at a seminar in Haifa, he quipped that this was the first time Cane and Abell had been together in the Holy Land for 5000 years. Having all four of the American chemists in the laboratory at the same time made for a great atmosphere—and we geneticists and microbiologists learned a lot about the mysteries of natural product chemistry from them as well.

After returning to the United States, Chaitan Khosla and his group, including his wife Susanne Ebert, whom he met and married in Norwich (Figure 8.13), made many

Figure 8.13. Chaitan Khosla and Susanne Ebert after their wedding in Norwich, June 1, 1991. (Courtesy of Tobias Kieser.)

combinations of genes from the actinorhodin, dihydrogranaticin, and tetracenomycin PKS gene clusters. They made a novel plasmid vector for constructing new combinations of the genes in *E. coli*; they then transferred the genes into an *S. coelicolor* strain which had had the whole actinorhodin cluster deleted so as to provide a host that could make no polyketides of its own to confuse the outcome of the experiments. The results were startling. Most of the hybrids made novel polyketides, and soon Chaitan's PhD student, Bob McDaniel, and his postdoctoral fellow, Hong Fu, had isolated and characterized a whole series of such unnatural natural products.

Molecules could now be designed on paper and created by engineering the hybrid strain, using PKS components from up to three *Streptomyces* species.[6] But this was largely based on empirical choices about which components to put together; the underlying programming rules were not apparent. The ketosynthase has to operate a specific number of times to build a carbon chain of the correct length, so how is this determined? How does the reductase know which keto-groups to modify? And how is the starter unit selected? These questions were not immediately answerable, even though the choices the PKS must make to build this family of "aromatic" polyketides (Figure 8.14) are rather limited: there is little variation in starter unit, almost all the extender units are malonyl-CoA, and only one or a few keto-groups are reduced. Paradoxically, the PKS enzymes that make more complex polyketides (Figure 8.15),

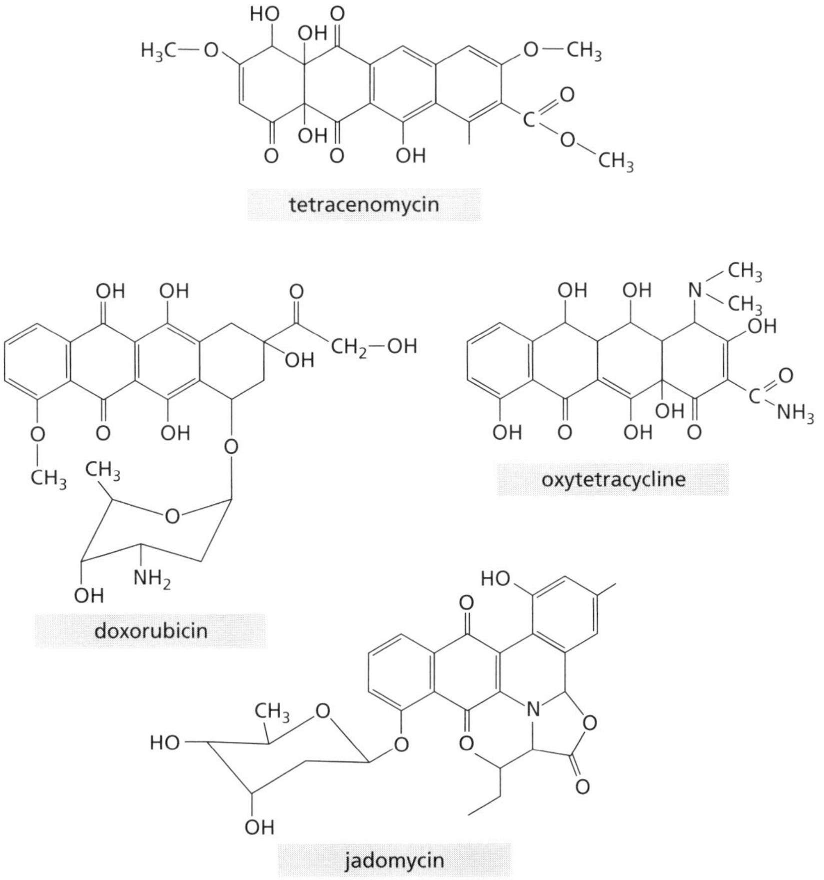

Figure 8.14. Some aromatic polyketides.

with many more choices in carbon chain assembly, turned out to be programmed in a completely logical way that could be identified just from the gene sequences encoding the enzymes. It was the elucidation of this programming logic that really set the new field of combinatorial biosynthesis of novel compounds on its way.

Hacking into the Enzymic Computer

The program for building complex polyketides was dramatically revealed through the work of Leonard Katz (Figure 8.16) and his team, including Stefano Donadio, at Abbott Laboratories in North Chicago, Illinois, and Peter Leadlay (Figure 8.12) and his group at the University of Cambridge, United Kingdom. They worked with the biosynthetic genes for erythromycin, a member of a very important group of polyketides called macrolides, in an actinomycete formerly classified as a *Streptomyces*

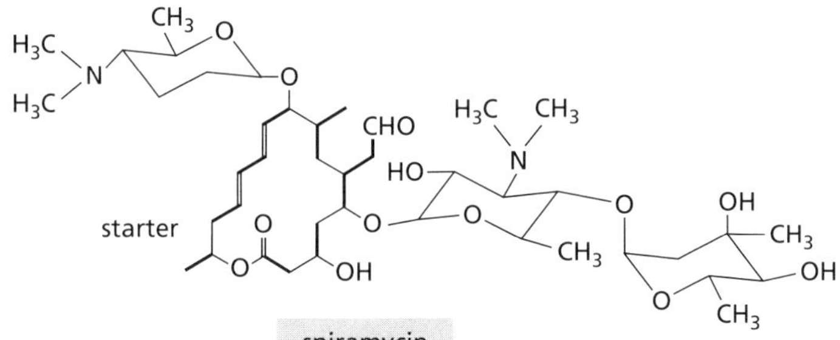

spiramycin
(SAAAAPBAA)

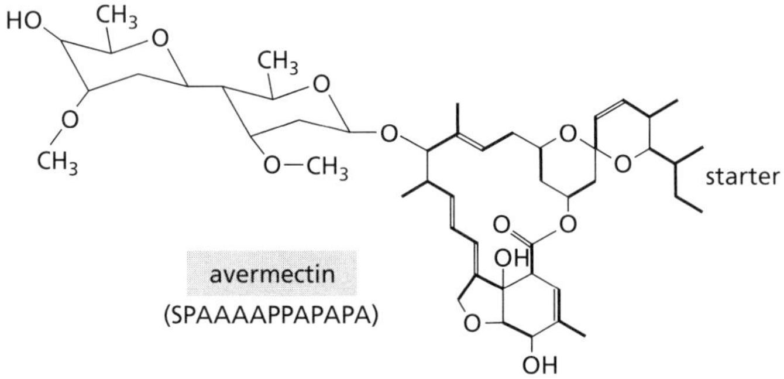

avermectin
(SPAAAAPPAPAPA)

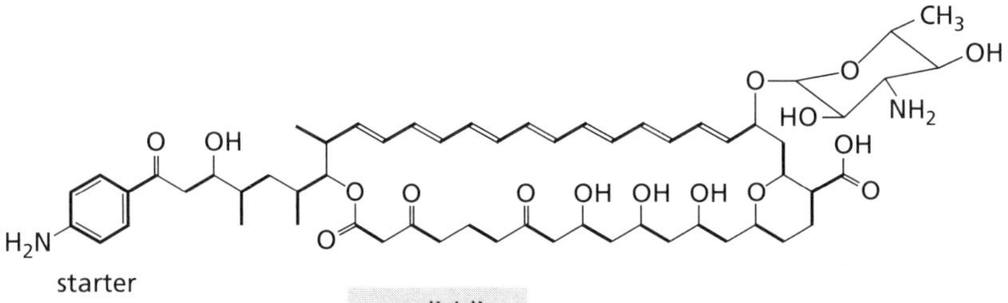

candicidin
(SAPPPAAAAAAAAPAAAAAAAA)

Figure 8.15. Some complex polyketides, each with one or more sugars attached. The code under each name represents the sequence of chain-building units: S, starter; A, acetate residue from incorporation of malonyl-CoA; P, propionate residue from incorporation of methylmalonyl-CoA; B, butyrate residue from incorporation of ethylmalonyl-CoA.

but now known as *Saccharopolyspora erythraea*. Erythromycin was one of Abbott's most important products. Both groups isolated and sequenced the DNA encoding the erythromycin PKS.[7] The results, published in 1990 and 1991, were extraordinary. Three enormous genes were revealed, each about 5000 base pairs long, compared with a typical bacterial gene of fewer than 1000 base pairs. Each gene encoded a protein carrying active site domains that corresponded to those on a vertebrate FAS and were even arranged in the same order.

Figure 8.16. Leonard Katz with Janet (Jan) Westpheling at the Society for Industrial Microbiology Annual Meeting, San Diego, July 2000. (Courtesy of Leonard Katz.).

The finding of a "vertebrate-type" enzyme in a bacterium was unexpected but not earth-shattering. The most remarkable discovery was that the enzyme consisted of six sets of active sites, or "modules," two on each protein. This was highly significant, because assembly of the carbon backbone of erythromycin requires six rounds of chain extension. A revolutionary model was suggested in which the carbon chain is built on a giant protein template that carries the six modules in order and functions like a car assembly line (Figure 8.17). Each module contains an acyl transferase domain to select extender units, a ketosynthase to make the carbon–carbon bond, and an ACP to carry the growing chain. The presence or absence of reductive domains in a module determines the modification of the corresponding keto-group. Modules 1, 2, 5, and 6 contain a ketoreductase but no dehydratase or enoyl reductase; therefore, a hydroxyl group is formed at the corresponding positions, with an appropriate stereochemistry each time, determined by that ketoreductase. Module 3 lacks any reductive domains, so the keto-group remains. Module 4 is the only one to

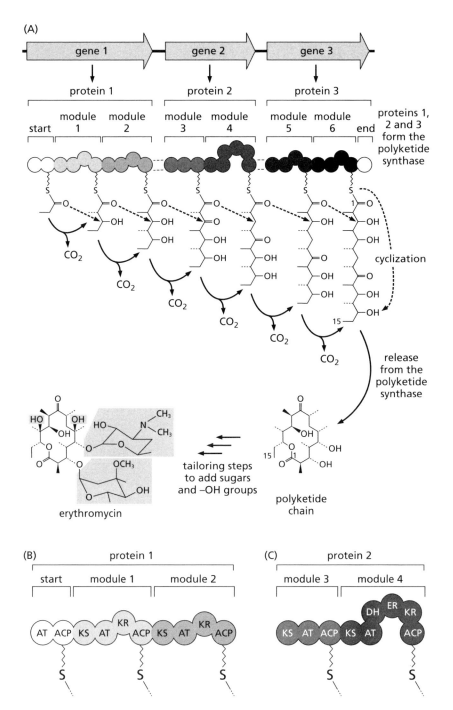

Figure 8.17. (A) The assembly line model for the erythromycin PKS. (B) and (C) Details of the domain structure of two of the PKS proteins. ACP, acyl carrier protein; AT, acyl transferase; DH, dehydratase; ER, enoyl reductase; KR, ketoreductase; KS, ketosynthase. The unshaded domain at the right end of protein 3 is the thioesterase.

contain the full suite of three domains, so the complete cycle of keto-group reduction occurs at that stage in carbon chain assembly. Two additional domains, an acyl transferase and an ACP, lie in front of the first module to select the starter unit and initiate carbon chain building, and there is a thioesterase domain after the last module to remove the completed chain from the enzyme by breaking the link with the tethering sulfur atom. The model for erythromycin biosynthesis was proposed from the amino acid sequence of the PKS, read from the DNA sequence of the genes, but soon the model was proved by inactivating particular reductive domains and observing predicted changes in the final product.

The simple programming logic of the erythromycin PKS offered a marvelous opportunity to make novel macrolides by rational genetic engineering. Camilla Kao (seen in Figure 9.8), a PhD student in Chaitan Khosla's laboratory at Stanford, took the first step. She transferred the genes for making the three giant PKS proteins on one piece of DNA from *Sac. erythraea* into the same *S. coelicolor* strain that had been used for the earlier work on mixing and matching genes for the actinorhodin-type PKS, whereupon it made the erythromycin polyketide.[8] Soon, using this experimentally amenable system, the Khosla group, in collaboration with Leonard Katz at Abbott and David Cane at Brown University, made many novel molecules by a series of domain and module swaps. Peter Leadlay's group in Cambridge made others. By comparing the sequences of acyl transferase domains from several other macrolide PKSs that had been sequenced, Peter could predict whether a given acyl transferase domain would select a malonyl extender unit or a more complicated unit such as methylmalonate (resulting in a methyl branch on the carbon chain, as in every chain-elongation step in erythromycin [Figure 8.17]) or one of several other alternatives in other macrolides.[9] Thus began what enthusiasts like me hope will be a second Golden Age of antibiotic discovery, in which useful designer polyketides are made by predictively inactivating, exchanging, or adding domains, groups of domains, or whole modules to the PKS. Whatever the long-term prospects, a new field of biotechnology has certainly opened up as the methodology is refined and the first fruits enter clinical trials.

Stanford University filed a series of patents on the technologies that Chaitan Khosla had developed, in the joint names of Stanford and the John Innes Centre. After a couple of false starts (the failing Exogene company was the first to license them), the patents were assigned to a new company called Kosan, cofounded by Chaitan Khosla and a professor at the University of California at San Francisco, Daniel Santi. Dan had experience with biotechnology start-ups, one successful (it was taken over by the large company Chiron) and others less so. His company, Parnassus, was the second licensee of the Stanford/John Innes patents but, before it had a chance to test the technology, it lost its venture capital support and had to close. Kosan began in Dan's house in San Francisco in 1995, with two former Parnassus staff, the chemist Gary Ashley and the molecular biologist Mary Betlach. The company rented a small space not far from Stanford, then moved to Burlingame, near the San Francisco International Airport, and later to larger premises at Hayward, just south of Oakland on the east side of San Francisco Bay, early in 1999 (Figure 8.18). By 2005, there were more than 130 employees of Kosan, including Leonard Katz and Dick Hutchinson. David Cane and I joined the Scientific Advisory Board.

Figure 8.18. Kosan Biosciences, Inc., in Hayward, California, with Daniel Santi, CEO, April 2005. (Courtesy of Anita Sharma, Kosan Biosciences.)

The company was founded on the ability to manipulate the choice of polyketide starter and extender units (controlled by the acyl transferase domains), the chain length (dictated by the number of modules), the ketoreductions (determined by the presence or absence of reduction/dehydration domains), and the stereochemistry of the hydroxyl groups and carbon side branches, as well as reactions after the building of the polyketide chain (e.g., addition of sugars). A series of exciting projects began at Kosan and at other organizations engaged in similar work, including a company in Cambridge founded by Peter Leadlay and his collaborator Jim Staunton, called Biotica. Various therapeutic goals that depend on the development of modified polyketides are being targeted, and hopefully there will be some medical and therefore commercial successes before long.

In a published example, the immunosuppressant polyketide FK520 was redesigned. This compound has a second activity—it helps nerves regenerate—but this cannot be exploited without first eliminating immunosuppression, because one would certainly not want to use an immunosuppressant in patients with no need for it. To act as an immunosuppressant, FK520 recognizes two receptors, an FK-binding protein and calcineurin, but only the interaction with the FK-binding protein is needed to promote nerve regeneration (Figure 8.19). The methoxy side chains ($-OCH_3$) at carbon atoms 13 and 15 of FK520 are needed for calcineurin binding. By engineering the molecule to have different carbon branches (or none) at these positions, a matter of genetically exchanging acyl transferase domains, Kosan made compounds that had the desired properties in animal models.[10]

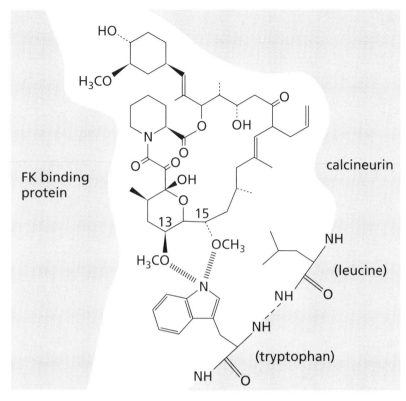

Figure 8.19. KF520 interacting with the FK binding protein and calcineurin. Cross-hatched lines show crucial hydrogen bonding between two methoxy groups (–OCH$_3$) on KF520 and a tryptophan residue on calcineurin. One of the methoxy groups also interacts with a leucine residue (no hydrogen bond).

In another project, Kosan made novel analogues of erythromycin called ketolides (so-called because they have a keto-group where the second sugar would normally be attached), which are important as antibacterial agents because they inhibit the ribosomes of pathogens that have become resistant to conventional erythromycin and other antibiotics. (The first such compound, telithromycin, was made by the Hoechst-Marion-Roussel Company, now part of Aventis, by chemical modification of erythromycin itself.) Kosan scientists made ketolides by a judicious marriage between synthetic chemistry and genetic engineering (Figure 8.20). First, a novel starter unit was synthesized and attached to a carrier molecule that mimics CoA but is small enough to enter cells. The artificial starter was fed to a strain of *S. coelicolor* containing an engineered erythromycin polyketide synthase that differs from the natural enzyme by having an inactive ketosynthase in module 1. This means that the fed molecule—a "diketide" analogue of the product that module 1 would normally make—skipped the first condensation and was accepted by module 2. The product moved down the PKS assembly line to end up as a novel polyketide with an interesting chemical group at the position where the normal starter would be located (carbon atom 15 in Figure 8.17). This compound was then fed to *Sac. erythraea* from which the erythromycin PKS genes, but not those for the erythromycin tailoring steps, had been deleted. This strain converted the novel polyketide into a novel erythromycin. Finally, some more synthetic chemistry produced potential ketolide drug candidates (Figure 8.20).[11]

Such metabolic and genetic juggling illustrates the possibilities of this new approach to drug discovery. But to widen the scope of making unnatural natural products, a new supply of biosynthetic genes will be needed to provide spare parts for engineering novel compounds, or to act as chemical scaffolds to be modified in predictable ways. Through the efforts of many laboratories over the past 10 years, 20 or 30 complete clusters for complex polyketides have been sequenced, but *Streptomyces* genomics is showing us that the known compounds represent only the tip of an iceberg.

Mining the Genomes

By analyzing the genome sequence of *S. coelicolor*, Greg Challis at the University of Warwick, a wizard at predicting possible biosynthetic pathways from gene se-

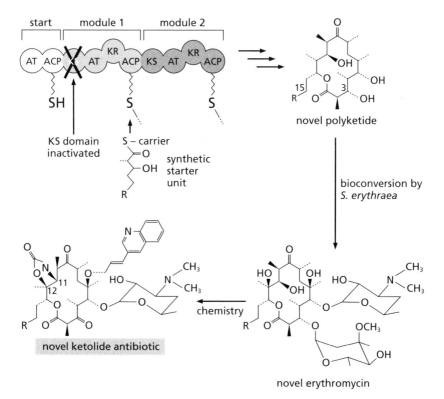

Figure 8.20. Production of a novel ketolide antibiotic by a combination of genetic engineering and chemical synthesis.

quences, discovered as many as two dozen gene clusters that are likely to make complex molecules with specialized functions, traditionally given the general name of secondary metabolites. They include the four known antibiotics, several complex lipids, the three iron-scavenging siderophores, geosmin, and many other, hypothetical, compounds. The *Streptomyces avermitilis* genome sequence contains more than 30 secondary metabolite gene clusters, most again making hitherto undetected compounds. The striking finding is that less than a quarter of the potential secondary metabolites of the two species are predicted to be the same, implying that the genomes of the actinomycetes, taken together, must contain the capacity to make a huge number of potentially interesting compounds, although this ability is often not expressed under typical conditions for screening biologically active agents. Perhaps this untapped resource could be "mined" by systematic sequencing of actinomycete genomes.

Ecopia Biosciences in Montreal, Canada, set out to do this. Their strategy was to sequence short genomic fragments, a few hundred base pairs long, that were cloned randomly from an actinomycete genome. By comparing the sequences with a database of more than 400 natural product gene clusters, they could identify promising clones and use them as probes to isolate larger clones of genomic DNA from the same organism for sequencing. In this way, complete biosynthetic gene clusters were identified and their possible products deduced. The predicted compounds are often not seen when the organism is grown in standard laboratory media but sometimes can be found under different conditions. A striking example is a set of strains that Ecopia, collaborating with Ben Shen at the University of Wisconsin, Madison, predicted from genome sequencing to make members of a class of amazingly potent antitumor agents called enediynes.[12] None of the strains at first produced them, but all did so when special growth conditions were employed. It is unlikely that they would have been found without prior knowledge from the genome sequence.

The idea that there are many useful natural products to be discovered was suggested in an interesting 2001 paper by Milind Watve and colleagues at Abasaheb Garware College, Pune, India.[13] They analyzed mathematically the numbers of *Streptomyces* antibiotics discovered over the decades and found that the curve fitted a model in which screening efforts are boosted by a previous year's success and the probability of finding a new antibiotic is a function of the fraction of antibiotics undiscovered so far. Extrapolating, they concluded that members of the genus *Streptomyces* alone are capable of producing at least 100,000 antibiotics, and they attributed much of the decline in the hit rate for finding new antibiotics to a reduction of screening efforts rather than an exhaustion of compounds.

Digging in the old literature, I came across a 1949 article by Albert Kelner, who worked at the famous Cold Spring Harbor Laboratory on Long Island, New York. He used ultraviolet and X-irradiation to induce mutations in *Streptomyces* with the objective of raising antibiotic productivity in the early days of strain improvement. In a separate experiment, he picked seven *Streptomyces* strains that had failed to reveal any antimicrobial activity. After irradiating them with X-rays, five gave rise to antibiotic-producing variants. Kelner commented: "Therefore, by genic manipulation of the cell we have a means for obtaining, in quantities sufficient for study, many of the metabolic products of the living organism that would otherwise be undetectable."[14] We can now interpret this prophetic statement in terms of the genetic potential of any randomly chosen streptomycete to make far more secondary metabolites than we usually detect. The new challenge is to understand the complex regulatory circuits that keep many of the secondary metabolic pathways "silent" until they are activated in response to specific signals, or derepressed by mutation, so as to switch on the genes in rational and predictable ways. Hopefully, there are general lessons to be learned that would allow the silent genes to be switched on without having to explore the physiology of every strain in detail.

A different approach is to bypass expression of genes in their native species and clone DNA fragments into a surrogate host in the hope that the genes will be relieved of repression and a product will appear. This concept has been combined with the realization that only a small percentage of microorganisms in natural habitats grow when environmental samples are cultured. To try to exploit the "viable but uncultivated" microbes, several companies are cloning DNA from the environment, even the entire DNA content of a soil or marine sample, into a convenient laboratory host. One of the first companies in this field was Terragen, set up by Julian Davies (Figure 8.21) just off the University of British Columbia campus in Vancouver, Canada, and later taken over by Cubist Pharmaceuticals, based in Lexington, Massachusetts. Another is the Diversa Corporation in San Diego, California. Time will tell whether this wildcard approach will pay off.

Meanwhile, the potential of *Streptomyces* genetic engineering to make novel therapeutics by working more predictably with gene clusters from known organisms, as practiced at Kosan and Biotica, is very exciting, but a prerequisite for commercial success is the ability to make enough of the engineered antibiotic, which is often produced in quantities even smaller than those of wild strains making their natural compounds. Can knowledge of *Streptomyces* genetics help with this, too?

Figure 8.21. Julian Davies at the Eighth International Symposium on Biology of Actinomycetes (ISBA'91), Madison, Wisconsin, August 1991. (Courtesy of Tobias Kieser, John Innes Centre.)

Improving Productivity

Many of the thousands of genes in a streptomycete affect the amount of antibiotic it makes. Impressive increases of productivity were achieved in mutation and screening programs, but they took many steps, each resulting in a small increment. Very recently, multiple rounds of protoplast fusion were shown to be a powerful way of bringing yield-enhancing mutations together to increase productivity in a much shorter time, as described in Chapter 4. Could knowledge of the genetics of antibiotic biosynthesis allow us to supplement such empirical approaches with rational engineering?

Streptomyces pristinaespiralis makes pristinamycin, a member of a family of antibiotics, each of which consists of two components, in this case called PI and PII, that act synergistically. Each compound alone inhibits bacteria without killing them, but together they kill bacteria. A problem faced by the Rhône-Poulenc Rorer group (now part of Aventis) and their academic collaborators at the University of Paris-Sud in developing pristinamycin as a drug was the poor water solubility of the PII antibiotic, so a water-soluble and therefore injectable derivative was made. In combination with a derivative of PI, the antibiotic was approved by the U.S. Food and Drug Administration (FDA) as Synercid (a 30:70 mixture of the PI and PII derivatives) to treat life-threatening infections by vancomycin-resistant *Enterococcus faecium* (Figure 8.22). But it is expensive to make.

The naturally produced PII is itself a mixture of two different chemical forms, PIIA and a useless precursor of it called PIIB; the precursor is converted to PIIA by a complex oxidation reaction catalyzed by two proteins that associate to form an active enzyme, but the reaction does not go to completion in the organism. The genes encoding the two proteins had been identified in the gene cluster for pristinamycin biosynthesis, so extra copies could be introduced into the industrial strain used to make pristinamycin. This was done by inserting the genes into a vector based on pSAM2, one of the plasmids that integrate stably into the chromosome in a transfer RNA gene (Chapter 4). The genes were expressed from a strong promoter that Mervyn Bibb had engineered from a natural promoter of *Sac. erythraea*, ensuring an abundant supply of the enzyme. The French group even locked the cloned genes into the chromosome by using a mutant version of the vector plasmid that carried the gene for the site-specific recombination enzyme needed to integrate the plasmid into the chromosome but lacked a second gene essential for the reverse process that loops out the plasmid. In this way, by using genetic technology from diverse sources, a metabolic bottleneck was overcome and a useful engineered strain was born.[15]

In a second example of rational metabolic engineering, scientists at Pfizer led by Kim Stutzman-Engwall (seen in Figure 9.8), in collaboration with Maxygen and its spinout Codexis in Redwood City, California,[16] used *in vitro* mutagenesis to improve industrial-scale production of Pfizer's successful Doramectin, their slice of a billion-dollar annual animal health market for avermectin derivatives as insecticides and nematocides. The starting strain made two compounds, of which only one, containing a double bond in the carbon chain, is useful. The gene encoding the enzyme that introduces the double bond was subjected to a process of *in vitro* DNA shuffling to improve the enzyme, and the mutated DNA was reintroduced into the starting strain to replace the wild-type gene. Screening only 28,000 colonies over four rounds of

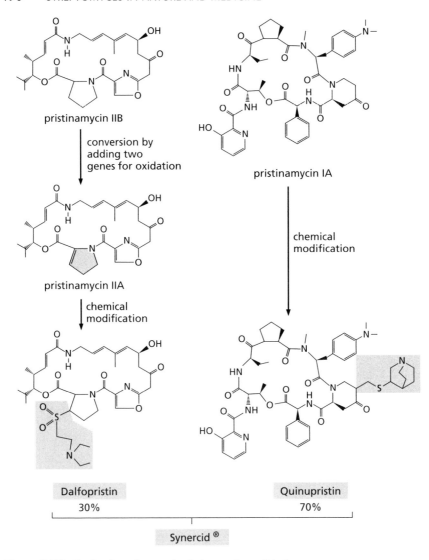

Figure 8.22. Production of a novel pristinamycin antibiotic.

mutagenesis yielded strains that made as much total avermectin as the parent, but with only 7% of the unwanted compound, a 23-fold improvement. The best mutants turned out to carry about 10 amino acid changes in the key enzyme, compared with the wild type. It is extremely unlikely that traditional *in vivo* mutagenesis and screening could have achieved the objective.

Improvement of pristinamycin and Doramectin production did not depend on a complete genome sequence for the antibiotic producer, but such information can help to improve strains such as *S. coelicolor*, one of the hosts used by Kosan for making unnatural natural products by genetic engineering. The Diversa Corporation sequenced the genome of a strain they call *S. diversa*,™ which they used for similar purposes. How could an inventory of the genetic potential of the host be used to optimize production of engineered molecules?

Figure 8.23 depicts a generalized biosynthetic pathway for converting a substrate to an antibiotic via several steps. Obviously, activity of the pathway enzymes is a prerequisite for antibiotic production, and a first objective for the genetic engineer might be to work on them. What concentrations of the enzymes are present? If the levels are too low, expressing the genes from stronger promoters may increase them. If we are working with an unnatural substrate and the enzymes have difficulty recognizing it, perhaps they can be mutated to improve their performance, as in the Doramectin example. But improving the pathway enzymes is only the beginning. If a natural substrate is being used, is it available in the cell in sufficient quantity? If not, perhaps we can increase expression of the genes that encode its biosynthesis. If the substrate is unnatural, it will have to be added to the growth medium. Will the cells take it up? If not, maybe we can boost its uptake. What about export of the unnatural end product from the cells? If this is sluggish, perhaps we can engineer a better transporter gene. Finally, if the desired end product of the pathway is an antibiotic—rather than a compound that will be chemically modified to make it into one— will the producing organism be resistant to it? If not, perhaps we can engineer the strain to tolerate the novel compound, perhaps by switching on the final steps of the pathway at a late stage of the fermentation, when the culture has stopped growing.

These are situations where we might want to accentuate gene expression, but in others we will need to inactivate genes that are having undesirable effects, such as encoding enzymes that divert the substrate or a biosynthetic intermediate into an unwanted shunt product. Inactivating them should yield more of the desired end product. Again, if we are using an unnatural substrate, the natural version will probably compete with it for the engineered pathway, because the biosynthetic enzymes

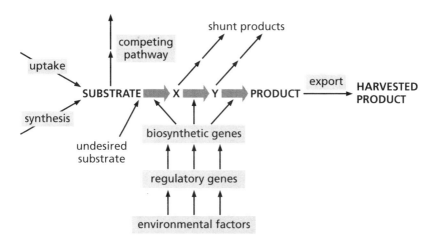

Figure 8.23. Some targets for improving antibiotic productivity. (Modified from a diagram supplied by Stephen del Cardayre, Codexis, Inc.)

have evolved to use the natural precursor efficiently. Perhaps we can inactivate the genes for biosynthesis of the natural substrate.

Adding or removing genes with obvious metabolic roles like these is already feasible in a host such as *S. coelicolor*, but rationally manipulating the many regulatory genes that influence the level of antibiotic production in response to environmental factors will require insight into the functions of much of the host's gene complement. In particular, we shall need to understand the roles of many of the genes in the organism that are currently labeled "of unknown function." A whole new field of "functional genomics" has grown up over the last few years to elucidate the expression and roles of all the genes in a sequenced genome. The next chapter describes how some of these techniques can be applied to *Streptomyces*.

9

Functional Genomics

The first complete bacterial genome sequence, for *Haemophilus influenzae*, was published in 1995,[1] followed the same year by the sequence for *Mycoplasma genitalium*. This simple, wall-less organism has one of the smallest genomes, only 580,070 base pairs and carrying just 470 genes. The genome sequence of a cyanobacterium (formerly called a blue-green alga) came in 1996, and in 1997 those for three chief model microbes: *Escherichia coli*, the workhorse of molecular microbiology; *Bacillus subtilis*, the model gram-positive bacillus; and baker's yeast, *Saccharomyces cerevisiae*, a favorite for studying eukaryote molecular and cell biology. Completion of the yeast sequence was a triumph of organization because it resulted from a cooperation among 100 laboratories in Canada, Japan, the United States, and Europe, including a European Union consortium of laboratories from almost every member state. The article in *Nature* summarizing the results had what may be a record of 633 authors.[2] It was by far the largest genome to be completed, with more than 12 million base pairs of DNA distributed among 16 separate chromosomes, although encoding only about 6000 genes. This is far fewer than in a bacterial chromosome of the same size, because eukaryotic chromosomes contain long stretches of DNA that do not code for proteins. By early 2006, there were 316 complete genomes of bacteria alone, and another 933 were in progress.[3]

Completion of a genome sequence for one's favorite organism feels like an enormous step, but after the initial euphoria a sense of reality sets in. Of the 7825 genes predicted in the *Streptomyces coelicolor* chromosome, the great majority could be assigned no biological role just by comparing them with other sequenced genes. They are officially "of no known function," a bit like being stateless or of no fixed abode. We can guess what kind of task many of them might perform, such as sensing an aspect of the environment or aiding the import or export of molecules, but we have no inkling what that aspect is or what those molecules might be.

To overcome such deficiencies, molecular biologists have developed a toolbox of techniques, both computational (bioinformatics) and experimental, to squeeze biological information from DNA sequences: this is functional genomics. The computer programs and biochemical procedures can be applied to any organism with a sequenced genome, but when they require experimental manipulation of the organism the techniques must be tailored to the practical tricks available for each microbe. This chapter describes them as applied to *Streptomyces*.

Searching the Genome Sequence

A prerequisite for studying a genome sequence is a user-friendly database. It holds the primary sequence data, the annotation with the predicted positions of potential genes and their putative functions, and maps showing all the features of the genome, including genes, repeated sequences, and transposons. There are links to the public databases containing all known genome information, so that BLAST comparisons (Chapter 5) or other analyses can be performed at the click of a mouse. Above all, the database needs to be searchable, for key words in the annotation and strings of base pairs and amino acids in the DNA and its translated products. Govind Chandra, on the staff at the John Innes Centre, extended the earlier work of Andreas Wietzorrek to construct a database for the *S. coelicolor* genome called ScoDB. Just type "ScoDB" into your browser to find it.

Understanding the response of *S. coelicolor* to oxidative stress provides a good example of using a genomic database to help deduce gene functions. As described in Chapter 7, free radicals cause sigma-R to stop binding to its anti-sigma and join with the core RNA polymerase to transcribe an appropriate suite of genes. Which genes are they? Three had already been identified: the sigma-R gene itself and two others. The promoter sequences of these genes, at the -10 and -35 positions upstream of the start of transcription, resembled each other enough to deduce a provisional "consensus" sequence consisting of the bases common to the three sequences. Mark Paget found 60 matches to this sequence in the *S. coelicolor* genome. Almost half were within predicted genes or otherwise in unreasonable places for promoter sequences, but 31, in addition to the 3 already identified, were a typical distance upstream of genes.

Back in the laboratory, experiments were done to see if RNA transcripts arose from the predicted start point for each of the promoters, but only after an oxidative stress. Figure 9.1A describes the method. Samples of RNA are isolated from mycelium grown with or without exposure to a chemical that mimics oxidative stress. Meanwhile, a synthetic DNA "probe" is made for each gene, extending from a point upstream of the predicted transcription start site to a point downstream of it (we shall see later how this can easily be done using the polymerase chain reaction) and is radioactively labeled at one end. The probe DNA is made single-stranded and mixed with the RNA samples, whereupon any messenger RNA (mRNA) molecules corresponding in sequence to the DNA fragment hybridize to it, making double-stranded molecules consisting of one DNA and one RNA strand, with single-stranded tails at

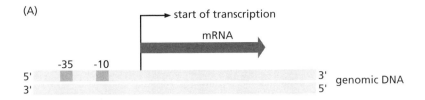

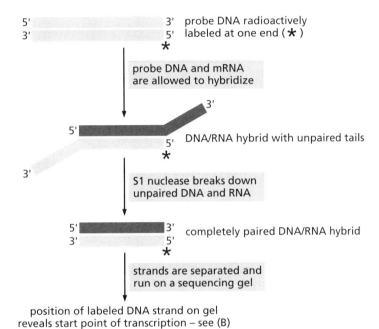

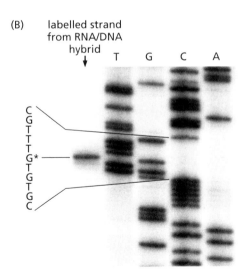

Figure 9.1. S1 nuclease mapping to identify a transcription start site. (A) Explanatory diagram. (B) Annotated gel with the DNA/RNA hybrid band deduced to start at a G (*) by reference to the tracks in a DNA sequencing gel. (Courtesy of Mark Buttner, John Innes Centre.)

either end. An enzyme called S1 nuclease, which digests single-stranded DNA or RNA but not double-stranded molecules, then removes the tails, trimming the hybrid molecules to the regions over which the RNA and DNA match. The hybrid molecules are detected as a band when the nucleic acid solution is run on a sequencing gel and exposed to X-ray film. The position of the band indicates the distance from the chosen downstream site to the start of transcription, and the sequence of the start site is read off from adjacent sequencing tracks in the gel (Figure 9.1B).

When Mark Paget did this for each of the 34 potential sigma-R–dependent promoters, 30 turned out to be real and active only in the mycelium exposed to oxidative stress. Some of the genes they controlled immediately made sense, such as the disulfide reductase that would convert potentially lethal S–S bridges back to free –SH groups, but most of them could not have been predicted. Now they could be studied as a group to see how they were involved in combating the dangers of oxidative stress. Because this information is likely to apply, with minor variations, to pathogenic actinomycetes like *Mycobacterium tuberculosis* and *Mycobacterium leprae*, the result could also aid research on these pathogens, which are harder to work with. A burst of oxidative stress is a defense the human immune system uses to kill invading bacteria, so understanding how the pathogen overcomes it may help in finding an effective antituberculosis drug (Chapter 10).

In this example, the genes of interest were predicted *in silico* to guide the subsequent experiments, but often there is no such guide, so two technologies have been devised to study expression of the whole complement of genes under specific conditions or in particular mutants. This sounds like an ambitious aim, and it is. It may well fall short of its target—there is no such thing as a complete understanding in science—but it can lead to rapid progress. One approach uses the set of mRNA transcripts to answer such questions as, Which genes are transcribed in certain tissues at a particular time in the developmental cycle, or in response to specific signals or stresses? The other technology looks at the complete set of proteins, asking similar questions about the translation of the genes. This may give a different answer, because some regulatory strategies control whether or not an mRNA is translated under a given set of conditions, rather than whether or not that mRNA is made in the first place. An example is the use of *bldA* as a regulator in *S. coelicolor* (Chapter 6): only when *bldA* transfer RNA (tRNA) is available can mRNA molecules containing the UUA codon (TTA in the DNA) be translated into protein.

Microarrays to Visualize the Transcriptome

The population of mRNA molecules in an organism or tissue is called the transcriptome, by analogy with the genome for the full set of genes. Transcriptome analysis depends on making DNA molecules corresponding to each and every gene and fixing them as spots in an ordered pattern, or array, on a surface. In the simplest version of the technique, this is a glass microscope slide measuring 7.5 × 2.5 cm (3 × 1 inch). When samples of mRNA from the target organism or tissue are then allowed to hybridize with the DNA spots, DNA–RNA hybrids are formed for all the genes transcribed under the conditions being studied. It sounds simple, but how is it achieved?

FUNCTIONAL GENOMICS   197

The DNA molecules for the array are made by a technique that has found many applications in studying DNA, not least in forensic investigations when only traces may be found at a crime scene, from blood, semen, or a single hair follicle. It depends on bulking up the DNA to make enough to work with, using a process of enzymic copying called the polymerase chain reaction, or PCR. Kary Mullis at the Cetus Corporation in Berkeley, California, the first biotech company to be founded, invented this technique in 1983, earning him a Nobel Prize for chemistry in 1993.[4] PCR is outlined in Figure 9.2. First, two short single-stranded DNA primers are made, approximately 20 bases long. They match sequences in the DNA to be amplified, pointing toward each

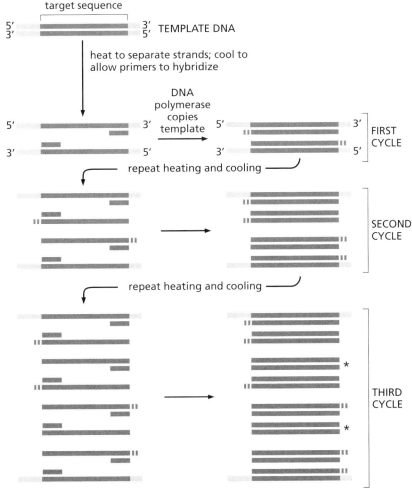

* asterisks show the first 'clean' copies of the target sequence
After 30 cycles there are over 1 billion copies of the target sequence

Figure 9.2. The polymerase chain reaction.

other and separated by a few hundred base pairs. Such primers are easy to synthesise chemically these days, and there are companies that make them to order.

A sample of genomic DNA, to be used as the template for amplification, is mixed with the pair of primers along with a DNA polymerase from a bacterium that lives in hot springs, so it works at a temperature that would destroy most enzymes. The double-stranded template DNA is made single-stranded by raising the temperature. The mixture is then cooled just enough for the primers to hybridize with their complementary sequences, one on each strand of the template DNA, whereupon the DNA polymerase copies the template to make double-stranded daughter molecules covering the distance between and including the primers. The temperature is raised again to separate the two strands, then lowered, and further molecules of the primers, which are present in excess, find their targets. A second cycle of DNA synthesis makes further copies of the desired sequence, and subsequent cycles increase the number exponentially. The required temperatures and the time spent at each are computer controlled, and many reactions are run in parallel on a robotic workstation.

The PCR products for all the genes are usually arranged in plastic trays with 96 wells, in 12 rows of 8. (These microtiter plates, measuring 12 × 8 cm, are the stock in trade of immunologists, who use them to test serial dilutions of antisera to estimate their ability to bind to a target; they are at the heart of many automated procedures in molecular biology.) Next comes the arraying process. A widely used system was invented by Pat Brown at Stanford University,[5] who provided a wonderful resource to the worldwide scientific community, not only about the science and results of arraying, but even posting on the Internet full do-it-yourself instructions for building an arraying machine.

A typical model has a head carrying 16 stainless steel tubes with specially shaped tips, like pen nibs. A robot is programmed to dip the set of tips into the first two rows of wells in a microtiter plate containing solutions of the PCR products; the tips take up just the right amount of liquid. The head then touches the tips on a specially coated glass slide, and they deposit 16 tiny spots of liquid, in the same pattern as the wells from which the DNA samples came. Without recharging the tips with liquid, the head moves to a second slide, and so on, until 250 have been printed. Then the robot rinses the tips, loads them with DNA from the next 16 wells, and prints another 16 spots on each slide, each just next to a spot in the first set. The process continues until all 96 samples from the first tray are printed, then moves to the next tray, and so on. After about 24 hours the whole genome is represented by a microscopic pattern of dried-down spots of DNA on each slide, almost 8000 of them for the *S. coelicolor* genome.

The printed arrays can be stored and used for many different experiments, and there is no need for each laboratory to prepare its own. Communities of scientists working on each of the most studied microbes have arranged for one laboratory to be funded to make the arrays for the whole community, or for a company to do it. (Colin Smith, who started as a PhD student with Keith Chater at the John Innes Centre elucidating the regulation of a primary metabolic pathway for glycerol metabolism in *S. coelicolor*, and who is now a professor at the University of Surrey, took on the task for the United Kingdom *Streptomyces* community and is now sending them all over the world.) Such collaborative arrangements have many benefits, not least because different groups of scientists can compare and integrate their results much better when they all use standard arrays.

Suppose we wanted to discover which genes are transcribed preferentially when *S. coelicolor* is exposed to a heat shock. We would isolate samples of mRNA from the mycelium exposed to, say, 42°C for a short time and from the control kept at 30°C. The mRNA is not used directly to hybridize to the DNA spots. Instead, a copy of the RNA is made in complementary DNA (cDNA) using reverse transcriptase—the enzyme that viruses such as HIV use to copy the RNA of the virus particle into the DNA that inserts itself into the host genome. The reaction mixture contains the four nucleotides needed to build DNA, with one carrying a fluorescent dye to tag the cDNA.

A green dye might be used for the control, and a red dye for the heat-shocked sample. Equal amounts of the two cDNA preparations are mixed and added to the arrayed DNA, whereupon the "green" and "red" molecules hybridize to each spot in proportion to their relative abundance. Excess cDNA is washed away, the slides are scanned in a microscope that measures the fluorescence of each spot at the two wavelengths, and a computer compares the amounts of green and red light. If the two are the same, the gene corresponding to that spot was transcribed to the same extent at the two temperatures, and the result is presented as a yellow dot in a magnified image of the array (Color Plate 9). If a gene was switched on preferentially at the higher temperature, the spot appears with a certain intensity of red. Genes that are less active at the higher temperature appear as green dots.

A nice early application of transcriptome analysis in *S. coelicolor* came from Stan Cohen's laboratory at Stanford. Jane Huang, a postdoctoral fellow there, made mRNA samples as the organism progressed from vegetative growth through to antibiotic production and compared message abundance for each gene at the later time points with the earliest sample from young vegetative mycelium.[6] One of her findings concerned the calcium-dependent antibiotic (CDA), a peptide made by the stepwise assembly of 11 amino acids into a chain in a process analogous to the way polyketides are made from organic acids (Chapter 8). The biosynthetic genes lie immediately next to a set of genes that would encode the enzymes for making tryptophan, an amino acid used at two points in building the CDA backbone. There is another set of tryptophan biosynthetic genes elsewhere in the genome, which the organism uses to make the tryptophan needed for protein synthesis during vegetative growth. Why should there be a second set, and is it significant that they lie next to the gene cluster for CDA biosynthesis?

Jane found that this set of tryptophan genes was transcribed only at the comparatively late stage in growth when CDA is made, and the time course of transcription was exactly in step with the antibiotic biosynthetic genes. This suggests that the organism uses a dedicated set of genes to make the tryptophan needed for antibiotic synthesis, which probably occurs in mycelium that has stopped growing and therefore no longer needs tryptophan for protein synthesis, another nice example of "tissue-specific" gene sets in *Streptomyces* (Chapter 6).

Seeing All the Proteins

Just as the set of mRNA molecules transcribed from a genome is called the transcriptome, a corresponding analysis of all the proteins tells us about the proteome. Typically this is done by making an extract from a sample of mycelium and separating the proteins according to their net electrical charge by running them along a strip of gel in an electric field. The strip is prepared in advance with a gradient of pH along its length, often from pH 4 to 11. Proteins take up characteristic positions along the gradient depending on their content of acidic amino acids, each of which adds a negative unit of charge to the protein, compared with the number of basic amino acids, which each add a positive charge. Next the strip is laid at one edge of a slab gel about 24 cm square, and the electric current draws the proteins into the slab under conditions in which they move at a rate depending on their size, the small ones fastest (Figure 9.3). When the gel is flooded with a protein stain, a complex pattern of spots is seen. Further resolution can be achieved by running "zoom" gels with the pH gradient in the strip gel covering various segments of the entire range, such as pH 4.5 to 5.5 in the example in Figure 9.4.

Patrick O'Farrell, while he was a graduate student at the University of Colorado, Boulder, worked out how to separate complex mixtures of proteins from a bacterium by this kind of two-dimensional gel electrophoresis in 1975.[7] What is new in the genomics era is the ability to relate each protein to the gene it comes from. This is done by cutting out a spot and digesting it with trypsin, which cuts protein chains into pieces wherever there is an arginine or a lysine in the amino acid sequence, yielding 15 to 20 fragments for an average-sized protein. The mixture is analyzed in a

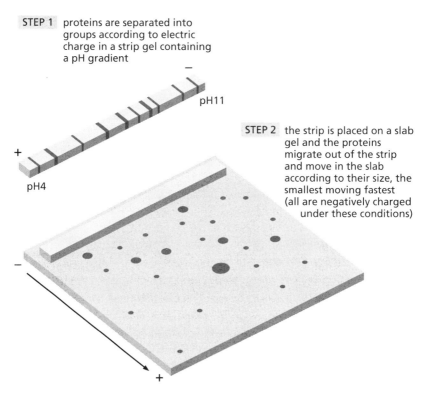

Figure 9.3. Separation of proteins by two-dimensional electrophoresis.

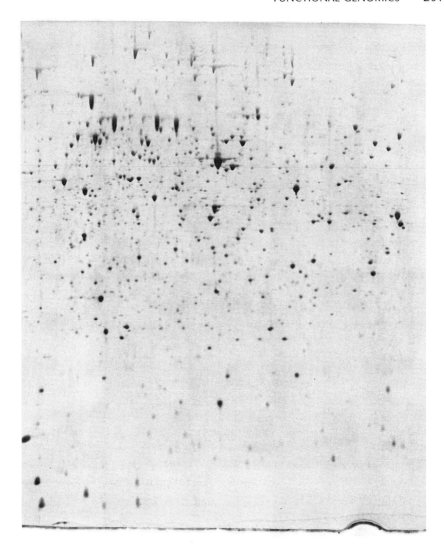

Figure 9.4. Analysis of *Streptomyces coelicolor* proteins by two-dimensional electrophoresis. This is a zoom gel with the narrow pH range of 4.5 (left) to 5.5 (right) in the first dimension (step 1 in Figure 9.3). Around 1500 protein spots are visible on the gel, of which over half could be identified by mass spectrometry. They range in molecular weight from about 150,000 at the top of the gel to 10,000 at the bottom. (Courtesy of Andrew Hesketh, John Innes Centre.)

mass spectrometer, which gives a precise value for the molecular mass of each fragment, reflecting its amino acid composition. With the aid of some fancy software, the result is compared with the pattern expected if each protein predicted from the genome sequence were digested with trypsin, and a unique match is made; no two proteins give precisely the same set of fragments.

The annotated set of spots can be compared with the set seen when the organism is grown under different conditions. Or we can compare the proteome in a mutant lacking a putative transcriptional activator with the wild type grown under the same conditions. Not all the proteins are seen, so information on the proteome is normally less complete than for the transcriptome. For example, regulatory proteins tend not to appear, because they are often present in only a few molecules per cell, and membrane proteins are difficult to extract in soluble form, but lots of useful information is obtained.

Analysis of the proteins can also tell us which of them are modified in response to a particular set of conditions, for example by adding a phosphate group, which increases the mass of a fragment by just 80 units. Protein phosphorylation is often a key step in a signaling pathway (Chapter 7). We may also be able to deduce the location of a protein by making separate protein extracts from the mycelial contents, or the walls, or the outside medium: many of the proteins will be found in one place only.

As for DNA arrays, a few laboratories can do a lot of groundwork to benefit the whole community. Each group needs to run its own gels, using proteins extracted from its own mutants, or under conditions of special interest to that laboratory. But posting results on the Internet allows the whole community to build on foundations laid by a few groups. Images of gels run under specific conditions can be animated so that clicking on a spot opens a file with the identity and properties of the protein. This was possible for more than 1000 spots in the *S. coelicolor* proteome in 2005, through the efforts of Andy Hesketh at the John Innes Centre in Norwich (Figure 9.5) and Charles Thompson and his group at the Biozentrum in Basel before he moved to the University of British Columbia in Vancouver in 2004.

Proteome analysis in *S. coelicolor* is already proving extremely useful. For example, the genome potentially encodes many hypothetical secondary metabolites. Are the genes expressed? Andy Hesketh has detected at least one protein from several of the gene clusters that Greg Challis predicted (Chapter 8), so the answer is

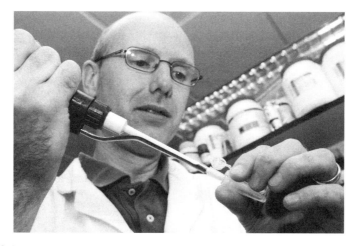

Figure 9.5. Andrew Hesketh at the John Innes Centre, April 2005. (Courtesy of Andrew Davis.)

"yes" for these clusters.[8] The hunt is now on for conditions under which the metabolites made by the pathways can be sought with the best chance of success.

Knocking Out the Genes

Transcriptomics and proteomics are powerful techniques for guiding further work. They immediately suggest hypotheses about the roles of genes, simply because they tend to be expressed together under a specific set of conditions, such as when antibiotic biosynthesis begins. Or the gene products appear or disappear together if a mutation is made in a repressor or activator gene. But is this a causal connection, or just a correlation? Studying mutations in genes is a classic way of answering such questions, but now, working from the genome sequence, we can inactivate the genes systematically and study the consequences.

There are two main ways of mutating all the genes in an organism. One depends on naturally occurring "jumping genes," or transposons, which inactivate genes by inserting into them. In the other, a stretch of DNA is artificially inserted into specific genes. Either way, if the gene is not essential under the conditions used, we obtain a mutant that can be examined to see what aspect of the normal metabolism or development has been changed. If we fail to isolate mutants, we deduce the gene to be essential.

Transposons occur in all forms of life. Barbara McClintock, who worked at the Cold Spring Harbor Laboratory on Long Island, New York, from 1941 until her death in 1992, and won the Nobel Prize for Physiology or Medicine in 1983, was the first to describe them, in maize. She first talked about them at the annual Cold Spring Harbor Symposium in 1951, but hardly anyone was prepared to accept that genes could move from one place to another on the chromosomes, and her breeding experiments were hard for the outsider to follow.[9] Her conclusions took a long time to be generally accepted, and then really only after transposons had been studied in bacteria at the end of the 1960s.[10] Direct analysis of the DNA, especially using techniques for visualizing plasmid DNA molecules in the electron microscope (Figure 9.6), put their existence beyond question.

It soon transpired that bacterial transposons were by no means an esoteric curiosity: they are responsible for the rapid spread of antibiotic resistance among pathogenic bacteria, which began to be a severe clinical problem even as the Golden Age of antibiotic discovery got under way (Chapter 2). Most horizontal transfer of antibiotic resistance between distantly related bacteria is due to plasmid transfer by mating, but any plasmid existing before the clinical use of antibiotics was unlikely to confer resistance to more than one chemical. The selective pressure imposed by the increasing use of antibiotics resulted in plasmids' acquiring multiple resistance genes. In the early 1960s, Tsutomu Watanabe at Keio University in Tokyo found that *Shigella* causing dysentery in Japan could transfer resistance to chloramphenicol, tetracycline, streptomycin, and a sulphonamide to other bacteria by conjugation.[11] Transposons had jumped one after the other onto the same plasmid. The consequences of such build-up of resistance to multiple antibiotics in dangerous pathogens such as methicillin-resistant *Staphylococcus aureus* (MRSA) are discussed in Chapter 2.

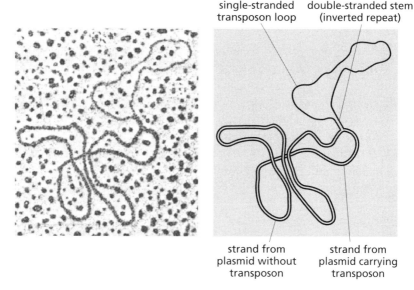

Figure 9.6. Visualizing a transposon by electron microscopy of DNA molecules. This is a hybrid molecule consisting of one strand from a plasmid carrying a transposon and the other from the transposon-free plasmid. Note the thick double-stranded DNA and the thinner, single-stranded DNA of the unpaired transposon. (Modified from an image supplied by Stanley Cohen, Stanford University.)

Figure 9.7 shows a generalized bacterial transposon and how it moves from one site on a plasmid or chromosome to another. Two features of the transposon are responsible. One consists of identical DNA sequences at either end of the transposon in opposite orientations, called terminal inverted repeats. The other is a gene encoding a transposase enzyme that recognizes these sequences and catalyzes a special nonhomologous recombination to move the transposon from its existing insertion site to a new one.

Shiau-Ta Chung at the Upjohn Company in Kalamazoo, Michigan, discovered a *Streptomyces fradiae* transposon in 1987. It did not confer antibiotic resistance, but Chung cloned a viomycin resistance gene into it to provide a selective marker gene. Kay Fowler, a PhD student with Tobias Kieser in Norwich (Figure 9.8), engineered this transposon into an excellent system for inactivating *S. coelicolor* genes.[12] She placed the transposon on a plasmid that replicates in *E. coli* and carries *oriT*, allowing it to be mobilized during mating into *Streptomyces* when the *E. coli* culture is mixed with *S. coelicolor* spores on an agar plate. The plasmid cannot replicate autonomously in *Streptomyces*, so only when the viomycin resistance gene has become part of the chromosome is the mycelium resistant to the antibiotic. When viomycin is poured over the plate (along with nalidixic acid to kill the *E. coli*), the colonies that appear arise from independent insertions of the transposon into the *S. coelicolor* chromosome.

Each of Kay Fowler's plates yielded about 1000 colonies. She collected spores from all of them to yield a pool of transposon insertions and made such pools from

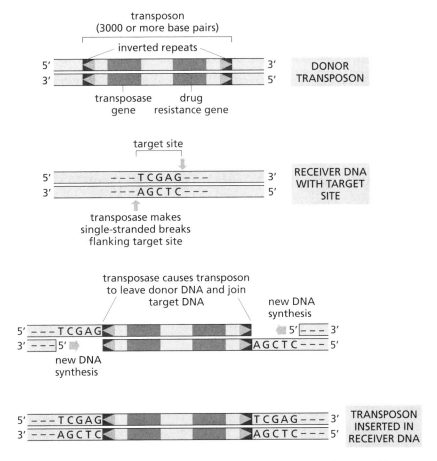

Figure 9.7. A generalized bacterial transposon and its mechanism of transposition.

100 different plates. She then mixed samples of these pools, 10 at a time, to yield 10 mixtures containing about 10,000 transposon mutants each, and finally she mixed these into a single pool of 100,000 mutants. Because the *S. coelicolor* genome contains 7825 predicted genes and 90% of the DNA represents genes, with only about 10% between genes, the final pool should contain on average more than 11 transposon insertions in every gene (90% of mutants are in genes, so 90,000 insertions are spread over 7825 genes, giving 11.5 insertions per gene). There would be no insertions in genes essential for growth under the conditions used, so every gene that could tolerate being inactivated would have even more insertions.

Next, Kay used PCR to identify mutants with the transposon inserted in particular genes. One of the primers, which was the same for all the genes, matched a sequence near one end of the transposon, pointing outward, and the other matched a sequence within the chosen gene (Figure 9.9). When the primers were used for PCR on DNA from the large pool as template, they hybridized to any DNA molecules

Figure 9.8. Tobias Kieser, with Kay Fowler (left) at a poster session at the 12th International Symposium on Biology of Actinomycetes (ISBA 2001), Vancouver, August 2001. The others are Camilla Kao of Stanford University and Kim Stutzman-Engwall of Pfizer, Inc. (extreme right). (Courtesy of Joyce Hopwood.)

with the transposon inserted in the target gene, and a PCR product was generated that could be detected as a band on a gel, indicating that the gene was not essential. Next, DNA from each of the intermediate pools was used as a PCR template, and an amplified band resulted from some of them. One of these was chosen, and the 10 small pools it contained were tested to find one that yielded a PCR product and therefore contained the desired insertion. The exact site of insertion was determined by sequencing the PCR product, revealing part of the transposon sequence and part of the target gene, with a boundary at the point at which the transposon was inserted. Because the pool was made from 1000 colonies, each with a different transposon insertion, the mutant of interest should represent 1 in every 1000 colonies when a sample of spores from the pool was plated out. It was then a question of identifying such a colony.

This was done by using a probe to find matching sequences in the set of colonies. The probe was the PCR product already sequenced. Kay made a print of many colonies on a filter membrane and broke the mycelium to release the DNA, which was made single-stranded in an alkaline solution so that it stuck to the filter. Next the filter was moistened with a solution containing the radioactive probe DNA, also made single-stranded. The unpaired probe DNA was rinsed away, and the filter was put in contact with X-ray film. All of the colonies contained the transposon sequence, because they all had transposon insertions somewhere in the genome, and all carried the target gene sequence, so the probe could in principle hybridize with DNA from every colony. However, only in the desired mutant would the two sequences be adjacent to each other, so only in that mutant would the entire length of the probe form

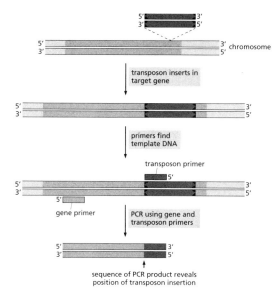

Figure 9.9. PCR to identify the position where a transposon has inserted into a target gene.

a completely double-stranded hybrid molecule with the chromosomal DNA (Figure 9.10A). Such hybrids remained intact at a high temperature, whereas the hybrids with unpaired "tails" separated, so only the desired colony yielded a dark spot on the X-ray film after a round of heating (Figure 9.10B).

Making random mutants like this with a transposon is a great way to find genes involved in a particular aspect of the biology of the organism, when it cannot be predicted which genes might be involved. The strategy is to pick mutants of the desired kind in the transposon library and use PCR as just described to find which genes contain the transposon. But for other objectives, we need to target chosen genes. For example, the working hypothesis is that all of the 49 ECF sigma factors (Chapter 7) respond to something going on outside the cell, but what might that be? One way to find out would be to knock them out one by one and see how the mutants respond to all of the possible conditions and stresses we can think of. And what about the thousands of genes of unknown function? An ambitious goal is to mutate each and every one of them: such a project for yeast involved a consortium of European Union laboratories, and one is now running in *S. coelicolor*.

For such targeted knockouts, a stretch of DNA is inserted artificially into each gene. Bertolt Gust, a postdoctoral fellow with Keith Chater in Norwich, devised an efficient way of doing this in *S. coelicolor*,[13] adapting technology first developed for *E. coli* by Kirill Datsenko and Barry Wanner at Purdue University, Indiana.[14] In fact, the genetic engineering itself is done in *E. coli,* and an engineered plasmid is then transferred by mating into *S. coelicolor*, where it disrupts the target gene. PCR is again the key. The template DNA for the PCR consists of an antibiotic resistance gene that works in both organisms, joined to *oriT* to mobilize DNA from *E. coli* into *S. coelicolor*. Two primers are made to match the ends of this template, and they also carry short tails corresponding to sequences near the ends of the gene to be dis-

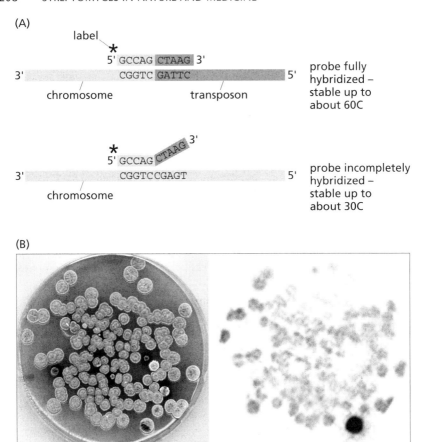

Figure 9.10. Colony hybridization to identify a transposon insertion. (A) A radioactively labeled DNA probe with sequences corresponding to the junction between a transposon and its target gene hybridizes over its entire length to the chromosome of the transposon mutant, but only over part of its length to the wild type with no transposon in that gene. (B) On the left is a plate of *Streptomyces* colonies to be screened for the transposon insertion, and on the right an autoradiograph with the dark image of the one colony in which the probe hybridizes over its full length at 60°C. (Courtesy of Kay Fowler, John Innes Centre.)

rupted, facing each other (Figure 9.11). After PCR, we have a stretch of DNA consisting of sequences from either end of the gene to be disrupted, with the resistance gene and *oriT* between them. The DNA is introduced by transformation into a special *E. coli* strain that is engineered to make homologous recombination very frequent and containing one of the cosmids that were used to sequence the *S. coelicolor* genome, the one carrying the desired gene. The introduced DNA fragment cannot replicate, so selection for antibiotic resistance produces a strain in which double crossing-over has replaced most of the functional gene on the cosmid by the resistance marker and *oriT*.

FUNCTIONAL GENOMICS

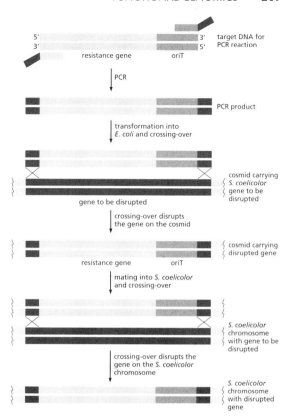

Figure 9.11. PCR-based gene disruption.

The *E. coli* strain is now mated with *S. coelicolor*, selecting for the antibiotic resistance gene and against the *E. coli* donor, yielding the result of another double crossing-over that replaces the resident copy of the target gene in the chromosome with the disrupted version. Many such experiments can be handled in parallel, so it is perfectly feasible to contemplate knocking out large numbers of genes. Soon, there should be libraries of mutant strains that can be made available to different laboratories to follow their own particular interests. Of course, if a gene is essential under the conditions used, no viable mutant can be made.

The ability to knock out every gene in an organism and study the consequences is ambitious but could be achieved relatively rapidly given enough commitment. But this would not be enough. Although some members of a family of genes, for example the ECF sigma factors, may have unique roles, there will certainly be cases where roles overlap, perhaps extensively or even completely. In these cases, knocking out each gene may hardly affect the phenotype: genes will need to be inactivated in pairs, or perhaps in larger numbers. So there will be no shortage of things to do for a long time to come.

This chapter has described a series of techniques devised to tackle the formidable task of elucidating the roles of the thousands of genes encoded in a whole-genome

DNA sequence. They have opened broad new vistas of knowledge that will keep communities of scientists busy in the short term and will undoubtedly provide a jumping-off point for imaginative and currently unpredictable leaps forward. I have shown how they are being applied to *Streptomyces,* but they are equally powerful wherever they are brought to bear, including on the pathogenic actinomycetes. The next chapter describes genetic work with the mycobacteria that cause tuberculosis and leprosy and how it can help in the effort to bring them under control.

10

Genomics Against Tuberculosis and Leprosy

The mycobacteria that cause tuberculosis and leprosy were some of the first actinomycetes to be described, in the 1870s and 1880s (Chapter 1). After World War II, streptomycin as a treatment for tuberculosis (TB) brought their relatives, the streptomycetes, to worldwide attention as antibiotic producers. In the following decades, actinomycete and fungal antibiotics helped to bring most bacterial infections under control, but those caused by *Mycobacterium tuberculosis* and *Mycobacterium leprae* remained very hard to treat. Much of this book has dealt with the genetic endowment of the streptomycetes and how it helps them meet the challenges of their soil environment. In this chapter, I describe how mycobacterial genetics and genomics is leading to a better understanding of the adaptations of *M. tuberculosis* and *M. leprae* to a very different life style, bringing nearer the prospect of eventually consigning TB and leprosy to history.

A Role For Streptomyces?

I gained a personal interest in the mycobacteria by accident. In December 1983, I was invited to the 15th International Congress of Genetics in New Delhi, India, to speak about our recent work on the genetics of antibiotic production by *Streptomyces*. The congress organizers paid for my travel out of a grant from the Special Programme for Research and Training in Tropical Diseases (TDR). Since 1975, the World Health Organization (WHO), the United Nations Development Programme, and the World Bank had funded TDR to help combat a series of tropical diseases that mainly affect poor populations. Malaria, leishmaniasis, bilharzia infections,

Chagas disease, and sleeping sickness are all caused by eukaryotic parasites, but there was one bacterial disease in the portfolio, leprosy.[1] In recognition of TDR's financial input to the congress, there was an informal session on Saturday afternoon about how genetics might help TDR's aims, and I was asked to be one of the speakers. Unsure how our work could be relevant, I talked about our recent successes in developing gene-cloning systems for *Streptomyces* and finding that genes from a range of bacteria were expressed in *Streptomyces* better than in many other hosts. Cloning *M. leprae* genes into a readily grown host might be a way to study them, side-stepping the problem of the pathogen's unculturability and leading to new knowledge that might be applied to combating the disease. Because the leprosy bacillus is an actinomycete, I suggested that its genes might be expressed particularly well in *Streptomyces*.

The session ended with only a short discussion. The following Tuesday, December 20, I was due to travel to Chandigarh, some 250 km (155 miles) north of Delhi and joint capital of the Indian states of Punjab and Haryana, to give a seminar on our antibiotic work at the recently founded Institute of Microbial Technology. My wife and I were picked up from our hotel at 7:00 AM for the drive to the institute, but the trip ended in disaster. Perhaps we should have been forewarned by the lethal reputation of the Grand Trunk Road connecting Delhi with the Punjab, with its litter of wrecked truck carcasses every few miles. At midday, in Ambala, about 45 km short of our destination, the taxi was pushed off the road while trying to overtake a truck, crashing into a concrete post. My wife broke her leg at the hip and was knocked unconscious, and I broke my arm. The driver and our guide, K. Lakshminarayana from Haryana University, were also unconscious (they came round some hours later).

My wife and I were rescued by a local bank manager, who happened to be passing in his car, and taken to the hospital, where my wife regained consciousness and her leg was put under traction by tying a rope around it, with a couple of bricks hanging over the end of the bed. We spent an eye-opening 6 hours there. Then, as a result of much-needed assistance from Lakshmi and Krishna Sagar, owners of the Oriental Science Apparatus Workshops across the road from the hospital, who had been asked to help us by the bank manager as Rotarian colleagues of his, we managed to get back to Delhi in an ambulance the Rotarians had donated to the city of Ambala. Through the good offices of the British Council, whose staff had coordinated my trip to the congress, we were admitted to a small hospital in the compound of the British High Commission. It was there that Indira Nath (Figure 10.1), an expert on leprosy at the All India Institute of Medical Sciences in Delhi, visited us a couple of days later. She had heard my talk at the congress but had had to rush away after the session and had not expected to meet me because we had been due to fly back to the United Kingdom the evening after the Chandigarh trip. She told me about leprosy, especially its complex immunology, all new to me, and mentioned researchers to contact after I got home. As a result, I came to know some visionary scientists, including the late Jo Colston at the National Institute for Medical Research in London, who continued my education about the mycobacteria, and the immunologist Barry Bloom from the Albert Einstein College of Medicine in New York, who was promoting WHO programs to harness the powers of molecular biology for research on TB and leprosy. We were awarded a small WHO grant to enable Tobias Kieser, collaborating with Jeremy Dale of Surrey University, to explore the idea I had floated at

Figure 10.1. Indira Nath at the All India Institute of Medical Sciences, New Delhi, middle to late 1970s. (Courtesy of Indira Nath.)

the Delhi congress of expressing *Mycobacterium* genes in *Streptomyces*, but mainly I would go to Geneva a couple of times a year between 1986 and 1991 to contribute to WHO discussions on the molecular biology of TB and leprosy and help evaluate grant applications.

Barry Bloom (Figure 10.2) was a dapper, charismatic character, combining scientific rigor with a burning desire to relieve the suffering caused by TB. He chaired many of the WHO meetings, and it was largely through his advocacy that several leading bacterial molecular biologists were persuaded to move into the difficult field

Figure 10.2. Barry Bloom at Harvard School of Public Health, where he became Dean in 1998. (© 2003 Rick Friedman, all rights reserved.)

of mycobacterial genetics, where they went on to make seminal contributions. Among many initiatives, funding for sequencing of the genomes of the tubercle and leprosy bacilli stemmed from discussions by the WHO committees.

The John Innes *Streptomyces* team helped to catalyze genetic studies on mycobacteria through one of a series of practical courses on *Streptomyces* genetic manipulation that we ran in the 1980s, supported by the European Molecular Biology Organisation (EMBO). European applicants planning to work with *Streptomyces* had priority on the courses, but in July 1985 we welcomed Bill Jacobs, an American who as a PhD student had pioneered expression of mycobacterium genes in *E. coli* and had then been invited to work with Barry Bloom to try to develop tools for genetically manipulating the mycobacteria themselves. He had told Barry that he would accept, provided he could first learn the tricks of manipulating *Streptomyces* protoplasts at John Innes. Bill found his visit "a most productive and life changing experience,"[2] and so did we. Big and bear-like, with mischievous eyes looking out from behind thick glasses (Figure 10.3), Bill was the life and soul of the two-week course: at the final party he gave an unforgettable rendering of the Beatles' *Yesterday* with words good-naturedly satirizing *Streptomyces* geneticists.

After his brief stay in Norwich, Bill made protoplasts of *Mycobacterium smegmatis*, a nonpathogenic model species and, using a precious batch of tested polyethylene glycol he had taken back with him, managed to introduce foreign DNA into a mycobacterium for the first time using the *Streptomyces* experiments as a model.[3] He went on to develop a whole series of powerful genetic techniques and vectors, including some based on phages that he isolated from the Bronx zoo, using the tricks he had picked up in the EMBO course. He generously acknowledges that learning about *Streptomyces* helped him in working with its cousin actinomycetes: "The development of genetic systems for mycobacteria clearly was a beneficiary of the *Streptomyces* influence, a reality for which I am eternally grateful."[2]

Figure 10.3. Bill Jacobs at Albert Einstein College of Medicine, New York. (© 1996 Kay Chernush.)

Tuberculosis and Leprosy, Diverse Diseases

TB and leprosy are very different diseases. Leprosy came to occupy a dreaded place in human society because it seemed so insidious. It developed slowly but led to grotesque disfigurement, thus acquiring a particular stigma: the very fact that a special noun, leper, evolved to describe a sufferer from the disease placed it apart from other afflictions. In early medieval England, dedicated lazar hospitals were opened on the edges of the main towns and cities, where lepers lived and were cared for. Norwich, the largest city in England in the 12th to 14th centuries (larger in area even than London), had many such establishments, at least five just outside some of the city gates (Figure 10.4).[4] The lepers were allowed out from the hospitals but were subject to a whole series of restrictions: they had to wear distinctive dress, with a bell or clapper on the belt; to touch goods they wanted to buy only with their staff; to drink only from their own cup; and to stand downwind of anyone they spoke with outside the hospital.

In spite of its unique status as a dreadful disease, leprosy never afflicted more than a tiny percentage of any population and did not cause epidemics. It was a constant presence in Europe until it slowly disappeared—in England no new lazar hospitals were founded after the 15th century, and the existing institutions gradually closed—but in warmer countries it remained a menace. Since the advent of effective drug therapy, the number of cases of leprosy registered for treatment worldwide has fallen dramatically, from almost 5.5 million in 1985 to just over one-tenth of that number by early 2000. Nevertheless, several hundreds of thousands of new cases are reported every year, and up to a million may be undetected. But one of the main issues in dealing with the disease today is to try to get the message across that leprosy is treatable and not a death sentence. I remember seeing a cartoon in *The Times of India* in the 1980s showing a patient having a consultation with a doctor. The caption read: "Thank goodness it's only leprosy" (presumably in contrast to a deadly cancer). And during a visit my wife and I made to the National Leprosy Control Centre in Sungei Buluh, Malaysia, in 1985, the Director, Dr. Lim Kuan Joo, made sure we touched some of the patients to impress on us that they were not to be feared. He emphasized that most of the inmates could safely go back to their villages after treatment, but many were not accepted. They therefore stayed indefinitely, and the colony of more than 1200 patients and almost 500 healthy staff, covering 560 acres (230 hectares), became largely self-sufficient, growing much of its own food as well as producing goods for sale.

TB is a very different matter. It was the White Plague, which killed millions down the centuries, first in Western Europe and later in North America, Africa, and Asia, after introduction into the indigenous populations.[5] It is still the number one human bacterial disease by far. WHO declared it a global emergency in 1993, the first disease to receive this dubious accolade. More than 2 million people die from TB every year—one every 10 seconds—and a staggering 2 billion people (one third of the world's population) have been infected at some time in their life. Many of these individuals harbor *M. tuberculosis* in a dormant state, and in up to a tenth of them (about 8 million people every year), the subclinical infection develops into active TB after years or even decades.

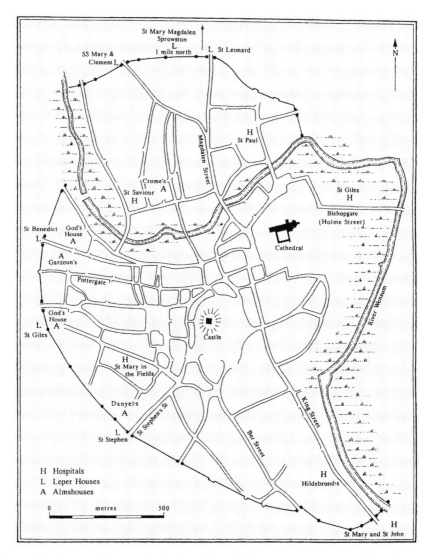

Figure 10.4. Map of the medieval city of Norwich showing the locations of leper houses outside the gates of the city, as well as hospitals for the sick poor and almshouses. (From Rawcliffe, C. and Wilson, R. [2004]. *Medieval Norwich.* London and New York: London & Hambledon.)

The extreme infectivity of TB distinguishes it from leprosy. Its spread does not depend on contact. *M. tuberculosis* is exhaled from the lungs of a person with active TB in a microscopic aerosol of liquid droplets, which quickly dry into tiny particles ideally adapted to be drawn into the airways of the lungs, where they deposit infectious bacilli in the alveoli, ready to invade the body of a healthy individual. This is why the highest incidence of TB is always in poor, cramped housing (it is especially rife today in Russian prisons). However, no section of the community escapes, and there is no shortage of celebrities who succumbed, including Cardinal Richelieu, Emily Brontë and her brother and all four sisters, John Keats, Frederick Chopin, Sir Walter Scott, Edgar Allan Poe, George Orwell, and Vivien Leigh,[6] as well as the wife and three sisters-in-law of Armauer Hansen, discoverer of the leprosy bacillus, and the wife of René Dubos, who discovered the first antibiotic (Chapter 1). In the 1930s, TB was still more often than not a death sentence, accounting for the extraordinary impact of the discovery in the mid-1940s that streptomycin could cure TB, even though, as we saw in Chapter 1, use of the drug had no big effect on overall mortality from the disease in developed nations, where it had been in decline for more than a century. It is now increasing again: by 130% between 1993 and 2003 among children in London, for example.[7]

One of the characteristics shared by *M. leprae* and *M. tuberculosis* is a slow growth rate. *M. tuberculosis* growing outside the body divides about once a day, so a single bacillus can develop into a visible colony on an agar plate in a few weeks, compared with one or two days for most bacteria. *M. leprae*, on the other hand, has still not been grown outside an animal, so all we have is an average estimate of the doubling time in the host, about 14 days.

The difficulty of finding an animal host for the leprosy bacillus in a situation approximating the human disease has hindered research and has an interesting history. By 1970, it was generally believed that *M. leprae* could grow only at a temperature below that of the human body. It seemed to prefer the cooler parts, such as the fingers and nose, and could produce a localized infection in mouse footpads, agreeing with the requirement for a temperature of less than 37°C. Therefore, when Eleanor Storrs, an expert on the nine-banded armadillo (Figure 10.5) at the Gulf South Research Institute at New Iberia, Louisiana, accompanied her husband W. F. Kirchheimer from the U.S. Public Health Service leprosy hospital at Carville, Louisiana, to a leprosy conference in the 1960s, she volunteered the information that this animal has a body temperature of 30°C to 36°C and might be a suitable host for *M. leprae*. Lo and behold, inoculation of the bacillus into armadillos resulted in a full-blown infection, and artificially infected armadillos came to supply almost all the bacteria needed in research laboratories. However, one of Nature's little twists was at work. It later turned out that armadillos have a peculiarity in their immune system that allows *M. leprae* to grow systemically; the low body temperature seems to be irrelevant.[8]

Like other pathogens, the bacteria causing leprosy and TB have a predilection for specific tissues and cell types. *M. leprae* enters the Schwann cells of the superficial peripheral nerves, preventing development of their normal protective coat, so that the nerves are destroyed by the immune system. The sensations of touch and heat are lost, and the sufferer neglects tissue damage that would normally heal if attended to. Leprosy varies from a very localized to a full-scale systemic infection with lots of bacteria in the circulatory system. At this end of the spectrum, deeper cutaneous trunk nerves are damaged, and the ensuing paralysis and osteoporosis lead to loss of fingers and toes.

TB, in contrast, develops when *M. tuberculosis* is taken up by the army of defensive white blood cells called macrophages that circulate around the body gobbling up bacterial invaders. Inside the macrophages, bacteria are engulfed in vacuoles called phagosomes that usually destroy them. The contents of the phagosome become acidic,

Figure 10.5. A stuffed nine-banded armadillo. This famous specimen belonged to Dick Rees, a pioneer of leprosy research at the National Institute for Medical Research, Mill Hill, London. The tail is jauntily arched above the body, whereas in life it trails behind like a mouse's tail. (Courtesy of the late Jo Colston.)

tending to prevent the bacteria from dividing. Then a component of the immune response kicks in and another kind of vacuole, the lysosome, fuses with the phagosome, delivering enzymes that digest the bacteria. *M. tuberculosis* can resist these attacks by interfering with acidification of the phagosomes and inhibiting their fusion with the lysosomes. In this way it cunningly "seduces" the macrophages, and many of them remain infected, sometimes to lead to the massive tissue destruction that is characteristic of TB. A more usual outcome, however, is a "tubercle," which consists of a central region of infected cells surrounded by uninfected phagocytes and so-called foamy giant cells, all sealed off by a fibrous capsule. Thus is established the latent state, which can trigger the disease in a poorly understood manner later in the patient's life.

Drug Treatment and Resistance

Although the discovery of streptomycin provided a breakthrough in the treatment of TB, and patients made previously unheard-of recoveries, the bacterium soon started to fight back. The first streptomycin-resistant *M. tuberculosis* appeared within 3 months after its large-scale introduction into medicine in 1946, by which time 80% of patients harbored at least a minority of resistant organisms. This is not surprising because of the need for a long period of treatment, which allows time for drug-resistant mutants to be selected in the patient's body. The answer was to use more than one antibiotic.⁹ Isoniazid, a synthetic chemical, was discovered in 1912 but found to be useful against TB only in 1952, when it started to be combined with streptomycin. The next major advance was another natural antibiotic, rifamycin, which was isolated from an actinomycete that has had several different names. First it was *Streptomyces mediterranei*, then *Nocardia mediterranea,* and is now *Amycolatopsis mediterranei*. Just as streptomycin kills bacteria selectively because its target is the bacterial ribosome, rifamycin inhibits the bacterial form of RNA polymerase but not the host enzyme. It was introduced into therapy, as a chemical derivative called rifampicin, in the 1960s in combination with streptomycin and isoniazid, and the three-drug cocktail cured almost all patients in 9 months, half the time needed for the earlier treatment, but still a very long period. In the 1980s, pyrazinamide was added, and the treatment regimen was shortened to 6 months. Then streptomycin was replaced by another synthetic drug, ethambutol, avoiding a problem with streptomycin, which caused patients treated in the early days to go deaf—a much better outcome than dying of TB but no longer an acceptable side effect. Within a decade, multidrug-resistant TB (MDR-TB) emerged and is now spreading fast.

MDR-TB is defined as disease that is resistant to at least isoniazid and rifampicin, but some strains can tolerate half a dozen drugs. Interestingly, resistance in *M. tuberculosis* has resulted entirely from mutations in genes encoding the drug targets—a ribosomal component for streptomycin, RNA polymerase for rifampicin, and so on—that render them insensitive to inhibition. This contrasts sharply with most bacterial pathogens, in which resistance genes on plasmids or phages are acquired from bacteria already harboring them. Obviously, MDR-TB is much harder to treat than TB caused by fully sensitive strains, and it needs a cocktail of up to seven drugs. Even then, half the patients may die from the disease. The situation for patients with HIV/AIDS is dire because the immune system, which otherwise can eliminate most of the invading bacteria when they are weakened by drug therapy, is overwhelmed. It has been estimated that a quarter of the recent increase in TB involves co-infection with HIV, which raises the chance of disease from an *M. tuberculosis* infection by a factor of 30, and when the *M. tuberculosis* is MDR there is little hope of a cure.¹⁰

Drug treatment for leprosy has followed a much smoother course, mainly because significant resistance has not developed. The first effective drug was dapsone, a synthetic chemical that inhibits a step in bacterial folic acid synthesis and thereby deprives *M. leprae* of an essential vitamin, but the introduction of rifampicin transformed the outcome of treatment. The two drugs taken together can usually wipe out the infection over a 6- to 12-month period, depending on the type of case. Those with large circulating populations of bacilli take the longer period and need three drugs: dapsone, rifampicin, and a synthetic drug called clofazimine.

The BCG Vaccine

If treatment is always long, resulting in huge problems of noncompliance and drug resistance, what about prevention? In view of the importance of these diseases, why is there no universally effective vaccine for TB or leprosy? Vaccines typically stimulate the immune system to produce proteins called antibodies, which circulate in the blood and lymph and recognize the invader in a highly specific manner, triggering its destruction by phagocytes. The trouble with TB and leprosy is that bacteria circulating in the body are not the main cause of the tissue damage that constitutes the diseases. Instead, it is the parasites insidiously developing inside host cells that need to be combated, in a process called cell-mediated immunity, and this is much harder to confer with a vaccine. Nevertheless, vaccination against TB is one of the world's most famous examples of protection from a bacterial disease, even if it is far from complete. It uses an agent called BCG, which stands for bacille Calmette-Guérin (Calmette and Guérin's bacillus).

Albert Calmette and Camille Guérin worked at the Pasteur Institute in Lille, a separate institution from the famous Paris institute, which was set up to prepare vaccines and antitoxins using horses, snakes, and other animals. They used *Mycobacterium bovis*, which causes TB in cattle but is responsible for about 5% of human cases. Starting in 1908, they subcultured the organism for as many as 231 generations on a laboratory medium and ended up with a strain that could still grow for a while when inoculated into the body (otherwise it could not have stimulated cell-mediated immunity) but caused only a mild infection that soon subsided. It was first used in humans in 1921 and since then has been given to more than 3 billion people, including 100 million newborn babies and almost 300 million adults every year, proving to be just about the safest vaccine ever made.

Reading one of Calmette and Guérin's papers, written in 1920 when they published their first findings on the protection of cattle from TB, throws fascinating light on some of the problems they had to surmount.[11] For one, they wanted to test its effectiveness against natural infection, not against artificially administered bacilli. They therefore tethered five tuberculous cows in a byre in front of and slightly higher than 10 disease-free heifers—five controls and five inoculated once, twice, or three times with BCG—so that the latter were constantly exposed to the feces and urine of the diseased cows. The experiments began in November 1912 but were severely hindered by the German invasion of northern France in 1914. First a curfew interfered with the experiments and then, in August 1915, they were "brutally interrupted" when the army requisitioned all cattle under pain of severe punishment. Not to be outdone, the two French doctors took the risk of clandestinely slaughtering their animals and performing autopsies to ascertain the extent of any TB. The animals that had received at least two doses of BCG were clear.

BCG has had a strange history as a vaccine. In several trials, it protected well against leprosy, an unexpected bonus, but the results with TB have been contradictory. A major trial in the United Kingdom found it highly effective, and BCG vaccination was introduced in 1954 for large numbers of children at most risk from catching TB in the major cities. In the United States, one trial showed similar results, but another had only a low success rate, so large-scale vaccination was not adopted. In Malawi no effect was seen, and in South India there was even a slight increase in TB in the inoculated group compared with the uninoculated controls.[12] There is no complete explanation, but there is a fair degree of consensus that BCG fails to establish protective immunity when the population has already been exposed, as in India, to a high level of infection by bacteria belonging to one of several *Mycobacterium* species that are constant, usually harmless, fellow-travelers of humans.[13]

Mycobacterial Genomics

The whole field of mycobacterial research took a quantum leap forward with the publication of the complete genome sequence of a representative of each of the two species, *M. tuberculosis* in 1998 and *M. leprae* in 2001. The TB strain was a famous culture called H37Rv, isolated in 1905 and used in laboratories around the world; bacteria isolated from an armadillo inoculated with *M. leprae* from an Indian patient provided the DNA for the *M. leprae* sequence. Sequencing and annotation were done at the Sanger Institute, with biological know-how from the TB and leprosy research communities. Stewart Cole, a Welshman who works at the Pasteur Institute in Paris, was a key player in both projects.

Both genomes, unlike linear *Streptomyces* chromosomes, are circular DNA molecules and thus resemble the genomes of the great majority of bacteria. The *M. tuberculosis* chromosome is 4,411,529 base pairs long and that of *M. leprae* 3,268,203.[14] No one was surprised that the leprosy bacillus had a smaller chromosome than the TB organism and so could carry fewer genes, because *M. leprae* is clearly a more specialized beast, and its inability to grow in culture probably reflects loss of genes needed for life on its own. The big surprise was that the loss of information is much greater than indicated just by the smaller size of the *M. leprae* genome. Whereas *M. tuberculosis* has approximately 4000 genes, *M. leprae* appears to have only about 1600 coding sequences, half the number expected for a genome of its size. About a quarter of the genome is made up of pseudogenes—damaged versions of genes found in active form in other bacteria—and another quarter does not contain recognizable genes at all and may be occupied mostly by gene remnants mutated beyond recognition. Therefore, the genome of the leprosy bacillus has suffered a massive downsizing in its coding capacity since it diverged from an ancestor common to all the mycobacteria.

The DNA sequences of the two pathogens are illuminating many aspects of *Mycobacterium* biology, both fundamental and applied. One interesting study compared the sequences of genes and other DNA features in hundreds of strains of the "*M. tuberculosis* complex," which includes *M. bovis* and a strain isolated from voles, as well as *M. tuberculosis*. The deduced rate of sequence change suggested that human TB is caused by descendants of a strain that developed some 15,000 to 20,000 years ago,[15] tantalizingly suggesting that this coincided with the domestication of cattle. However, the implication that *M. tuberculosis* evolved from the cattle pathogen, *M. bovis*, is not supported because the human strain has a larger, not a smaller, genome than *M. bovis*, which has undergone a whole series of deletions relative to *M. tuberculosis* in adapting to infect many different mammals.[16] However, it is still possible that *M. tuberculosis* evolved into a strictly human pathogen when people began to live together in relatively crowded conditions, rather than in small, widely dispersed groups.

Another detailed study, comparing BCG strains kept in different laboratories for decades, has thrown light on the events that took place as Calmette and Guérin repeatedly subcultured *M. bovis* on laboratory media, and others continued for another 1000 generations in laboratories around the world between 1920 and 1960—when it first became possible to freeze-dry the bacteria and thereby avoid the need for constant subculture.[17] Many genes in several clusters in virulent *M. bovis* are missing from some BCG strains, and all lack one cluster, which presumably was lost during the original attenuation process. The later deletions may have been undesirable, perhaps explaining in part why the vaccine does not always take as well as it should. Maybe there was automatic selection for better growth on laboratory media, correlated with less vigorous growth in the body. Indeed, the genes deleted since 1920 include a disproportionate number that would encode transcriptional regulators—exactly the kinds of genes that might be involved in responding to changes in the body by switching on or off appropriate sets of genes, just as *Streptomyces* monitors changes in its soil environment.

Possible applications of such genomic comparisons include TB diagnosis and the tailoring of a superior vaccine strain by exploiting differences between the BCG sequence and that of virulent *M. tuberculosis*. Various BCG strains lack up to 100 genes found in the H37Rv strain, and some might be used to diagnose TB.[17] The standard test depends on skin inoculation with an extract of tubercle bacilli called tuberculin, developed by Robert Koch as a TB treatment. It failed as a treatment, but proved successful in diagnosis. The problem is that BCG leads to the same inflammatory reaction to tuberculin as TB infection, making the diagnosis of active TB in vaccinated people very difficult. Could a new diagnostic material be made using only proteins expressed from cloned genes present in virulent *M. tuberculosis* and absent from the BCG genome? Or could a vaccine be made from cloned *M. tuberculosis*-specific proteins (a so-called subunit vaccine), followed by a booster inoculation with BCG, which seems to be uniquely able to trigger a cell-mediated immune response?

Clues to the basis of virulence have not yet emerged from the mycobacterial genome sequences. One of the most important concepts coming from genome sequencing of other pathogens has been "pathogenicity islands."[18] Disease-causing bacteria often have chromosomal DNA sequences that are very similar, overall, to those of their nonpathogenic relatives except for extra groups of genes (the pathogenicity islands) that encode proteins needed for some aspect of the infection process, such as a toxin, an enzyme, or a receptor to attach to or invade host cells. A nonpathogenic *Escherichia coli* strain can cause human disease by acquiring different sets of pathogenicity genes. In one example, a 35–kb segment of DNA has become inserted into a transfer RNA gene[19] in a process reminiscent of the way integrating plasmids insert into the *Streptomyces* chromosome. The segment carries genes for proteins that attach to and modify the lining of the small intestine, making it a suitable niche for the bacteria to grow, where they cause severe diarrhea that can be fatal in young children. A similar set of genes is found in other gut pathogens, and it seems that they have comparatively recently moved between unrelated bacteria by horizontal gene transfer.

Other *E. coli* strains cause infections of the urinary tract and have a different set of virulence genes on a 70-kb stretch of DNA, but it is inserted at precisely the same spot on the host chromosome as the smaller segment in the diarrhea-causing strains.[19] The genome sequence of *E. coli* O157:H7, which periodically causes fatal outbreaks of food poisoning when beef is contaminated with the intestinal contents of slaughtered cattle, revealed 3574 genes, of which as many as 1387 are different from those of the laboratory strain used for genetic work and are in islands of various sizes.[20] The concept of "plug-in" islands of genes that convert nonpathogens into pathogens is not new. For example, it has been known for more than 50 years that a gene on a prophage inserted into the bacterial genome encodes the lethal toxin produced by the diphtheria bacillus, another actinomycete distantly related to mycobacteria. Only recently have bacterial genome sequences shown that horizontal gene transfer is one of the most important drivers of pathogen evolution.

A few streptomycetes are plant pathogens. The best known is *Streptomyces scabies*, which causes potato scab. This disfigures the tubers, rather than destroying them, but because it can significantly depress the price it is nevertheless serious for those who make their living by growing potatoes. The hyphae of the pathogen enter the tubers through the lenticels, pores in the skin that allow the potato cells to breathe. This leads to a tumor-like growth that bursts the potato skin, which peels back in a characteristic way (Color Plate 10). Although all potato scab organisms were originally named *S. scabies* simply because they cause the disease, it is now clear that several quite distantly related species have acquired the ability to infect potatoes. Rosemary Loria and her group at Cornell University in Ithaca, New York, found that they harbor a large pathogenicity island carrying genes needed to cause scab, including those for biosynthesis of a tumor-inducing toxin called thaxtomin, and that the island could be transferred by mating to a nonpathogenic *Streptomyces* strain, whereupon it integrated into the genome and conferred pathogenicity.[21]

Remarkably, no such virulence factor–encoding islands have been unambiguously identified in the genome of either the tubercle or the leprosy bacillus, but perhaps this will change as more mycobacterial sequences become available. Meanwhile, several individual components of virulence have been identified. A hypothesis that fatty compounds coating the surface of *M. tuberculosis* (responsible for Robert Koch's difficulty in staining it [Chapter 1]) are involved in its ability to avoid intracellular killing has been supported strongly: more than 20% of its genome encodes enzymes to build fatty acids. Bill Jacobs' group in New York identified the targets for pyrazinamide and isoniazid as two of the fatty acid synthases of the bacterium.[22] The genome sequence also predicts more than 150 unusual proteins with repeated runs of the same amino acids. They too are probably components of the cell membrane or wall and could be protective.[23]

The genome sequences also suggest that both pathogens obtain energy mainly by breaking down host lipids. *M. tuberculosis* has a vast armory of genes to encode the necessary enzymes, and even *M. leprae* has kept them, while losing enzymes that most bacteria use to feed on a wide range of other materials. But the leprosy bacillus has retained a virtually full set of genes for making its own amino acids, vitamins, and other molecules that all cells need, whereas many pathogens that inhabit blood or lung fluid have lost them. Presumably these compounds are not available inside the cells the leprosy bacillus infects. Unfortunately, the genome sequence does not immediately answer the perennial question as to why *M. leprae* cannot be cultured, but clues must be waiting there to be interpreted.

For both TB and leprosy, the timely identification of infected individuals is crucial to therapy and control of the disease, not only because of the benefit to patients of treatment before serious tissue damage occurs, but also to reduce the number of others they can infect. DNA analysis allows this and, through the polymerase chain reaction (PCR), the difference in sequence between a sensitive and a resistant drug target can be detected in 24 to 48 hours, instead of the weeks needed to culture TB bacteria from tissue or sputum samples and test them for drug susceptibility.[24] These are significant advances. However, the big hope remains the development of new, more rapidly effective treatments for TB. What progress is being made in harnessing genomics to discover promising new targets and the drugs to interact with them?

New Drugs From Genomics?

Known antibiotics inhibit a restricted range of cellular functions. In order to find new targets, pharmaceutical companies have turned to genomic sequencing of pathogens, either commissioning private sequences from commercial enterprises or working with sequences freely available from public institutions, of which the Sanger Institute is a major player. The complete inventory of a pathogen's predicted proteins is scanned to find potential targets fitting a group of criteria. First, the targets should not have a close relative in the human genome, so as to minimize the chance that interacting with the target would be toxic to the host. Second, targets should be present in a suitable range of pathogens, depending on the diseases to be treated. Third, the target proteins should be essential for the survival of the pathogen and should be expressed inside the host, not merely when the pathogen grows on an artificial medium.

It is early days for the *M. tuberculosis* genome sequence to have yielded definitive targets for drug development, but promising candidates are being reported at an ever-increasing rate. Of course to find and license a good drug to inhibit such targets will take a lot longer—the recently established, nonprofit Global Alliance for TB Drug Development has set a date of 2010 to register the first new drug.[25] Transcriptome analysis is playing a part in target discovery. For example, Pat Brown, Gary Schoolnik, and their colleagues at Stanford University found genes for a biochemical pathway that were expressed only when isoniazid was present, including the gene encoding the known isoniazid target, making it highly likely that the newly identified members of the same pathway would also make promising targets.[26] Again, Eugenie Dubnau and Issar Smith at the Public Health Research Institute in Newark, New Jersey, identified a suite of genes that were switched on only inside macrophages, including genes involved in lipid metabolism. They also found an unexpected transcriptional regulator, a membrane pump, and genes likely to be involved in building the mycobacterial cell envelope.[27] And Thomas Shinnick's group at Emory University in Atlanta found genes that were switched on by the kind of acidic environment found in the phagosomes.[28] These again included genes involved in fatty acid synthesis, but also a gene encoding an enzyme called isocitrate lyase, which is needed to use acetate derived from fatty acid breakdown to make other essential molecules. This enzyme is a potential target for a drug that could combat *M. tuberculosis* in its dormant state, because Bill Jacobs and his colleagues have shown that a mutant with the gene inactivated can grow perfectly well in mice for the first 2 weeks of an acute infection but then rapidly disappears, while the wild type persists indefinitely.[29]

Comparative actinomycete genetics also may contribute to finding new mycobacterial targets to inhibit. Keith Chater's group in Norwich discovered that *Streptomyces coelicolor* and *Streptomyces lividans* contain half a dozen genes encoding proteins resembling the product of the *whiB* developmental regulator gene (Chapter 6); they presumably play subtly different roles in sporulation or other aspects of the organism's biology.[30] Multiple *whiB* homologues are confined to the actinomycetes but are found throughout the group, including the nonsporulating mycobacteria. That they play an important role in the biology of *M. tuberculosis* emerged from a recent finding by Charles Thompson's group in Basel. They had isolated an *S. lividans* mutant that was super-sensitive to a range of antibiotics that barely affect the wild type and found that the mutation was in one of the *whiB* homologues. The same phenotype was observed when the corresponding gene was mutated in *M. tuberculosis*, suggesting that a novel therapy might be developed by finding an inhibitor of the protein product of this gene and combining it with one or more already tried and tested antibiotics.[31]

Very recently, in a remarkable use of genomics, a totally new target for TB drug therapy was demonstrated.[32] The drug candidate itself, a promising member of the quinolone class of synthetic chemicals, was identified by the oldest method known—killing a culture of a *Mycobacterium*—but the target was found by whole genome sequencing. A group headed by Koen Andries of the Belgian unit of the pharmaceutical company Johnson and Johnson sequenced "to near-completion" the genomes of three mutants selected for resistance to the compound. Comparing each mutant sequence with the wild type, they found that the only base changes were in a gene for the membrane pump that generates energy by transporting hydrogen ions into the cell. Thus, genomics short-circuited what would have been a very long quest for the drug target by classic biochemical experiments.

This chapter has delivered a mixed message. On the one hand, TB is on the march again and will inflict many more casualties before it is finally brought under control. On the other, there are rays of hope coming from advances in immunology and cell biology, and especially from molecular biology and genomics. It will be exciting to follow the progress of this quest for a cure for one of mankind's greatest scourges.

Conclusion

"They are among the most beautiful, fascinating, and useful of microbes." This is how I described the actinomycetes in the opening paragraph of the book. How far have I justified these claims?

The varied colors of *Streptomyces* colonies seen with the naked eye certainly make them more attractive than the gray piles of cells typical of other bacteria, and examination of the colonies at every level of magnification reveals a complex and interesting architecture. But beauty is in the eye of the beholder, so individual readers will judge whether the actinomycetes portrayed in some of the photographs meet their criteria of visual beauty. While enjoying them, they can imagine the beautiful smell of the woods that streptomycetes bring to the microbiology laboratory.

I should be surprised if readers who get as far as Chapter 6 are not fascinated by the ways in which actinomycetes have adapted morphologically to life in the soil. New microscopical techniques for visualizing subcellular components are revealing a whole new world of understanding of the growth and reproduction of bacteria. Mostly invented in laboratories studying *Escherichia coli* and *Bacillus subtilis* over the past few years, these techniques label proteins with fluorescent tags via specific antibodies, or nucleic acids via hybridization probes. They are being used to track chromosome ends, origins of chromosome replication, or interaction points for the partition protein, but this is surely only the start of revealing the wonders of actinomycete microengineering as the organisms develop into colonies much more complex than those of other bacteria.

Streptomycetes are fascinating, too, in their physiological interactions with other organisms, microbial or macroscopic, competitors or colleagues. I touched on some of these relationships in Chapter 6, but again we are seeing only the beginning of what will surely develop into a much greater understanding as answers are found to many questions. It seems that the relationships between actinomycetes and other organisms, including higher plants and insects, are much more widespread than was thought just a few years ago. What are the signals that allow a streptomycete to sense a suitable plant or ant to colonize, and how does a soil invertebrate detect a *Streptomyces* colony and disperse its spores? What about the myriad secondary metabolites that genome sequencing reveals? What signals trigger their production, and what are their true ecological roles? If they are antibiotics, how did the producer manage to evolve a self-protective strategy in time to counter the potential lethality of the product? And if they are immunosuppressants, what has been the selective pressure for them to evolve?

Since actinomycete products account for about half of the $25 billion annual antibiotic sales, these bacteria have certainly been useful up to now, but will the success of antibiotics continue? Looking at the trend in natural product discovery over the decades (Figure 2.1), the decline after the 1960 peak is striking. Moreover, the relatively few natural compounds introduced into medicine or agriculture since 1975 have included only two anti-infectives, teicoplanin and thienamycin. The other compounds are anticancer agents, insecticides, immunosuppressants, and cholesterol-lowering compounds. At the same time, multidrug-resistant bacterial pathogens have been emerging at an ever-increasing rate, so there is certainly an ongoing need for new antibacterial antibiotics. Why is the need not being met more successfully?

It seems inescapable that the recent failure to introduce new anti-infective agents into medicine in part reflects the huge and escalating costs of the preclinical tests and clinical trials needed to bring new drugs to market in the face of increasingly stringent safety regulations and the hazards of unexpected side effects, which often become apparent only with the combined experience of large numbers of patients over several years. Not surprisingly, therefore, most big pharmaceutical companies have de-emphasized anti-infectives, which must now jump almost impossibly high regulatory hurdles, in favor of drugs that are easier to register. Anticancer agents will always show toxicity, so new compounds do not fail simply for that: the challenge is to improve on the therapeutic window of existing drugs, but without expecting to avoid all side effects. Other drugs, such as immunosuppressants, as well as statins to reduce cholesterol, will be taken by patients over much longer periods than antibacterials, even longer than anticancer drugs, so side effects are certainly a big issue here, but the promise of larger eventual returns to the successful manufacturer of a compound that patients may take for decades, rather than a week or so, will support very expensive research programs to try to come up with the ideal drug.

Tuberculosis (TB) might seem to offer an attractive target for new antibiotics because of the huge numbers of sufferers and the long period of treatment compared with other bacterial infections, but here there is undoubtedly an economic disincentive to big pharmaceutical companies to invest heavily in research. As with the well-publicized example of anti-AIDS medicines, sufferers in developing countries are mostly unable to pay the costs of bringing a TB drug to the clinic. The recent rise of multidrug-resistant TB in developed countries is an approaching cloud, but it might have a silver lining if it helps to motivate the pharmaceutical and biotech industries to put serious effort into combating the world's most serious bacterial disease. As I describe in Chapter 10, mycobacterial genomics promises to illuminate the basis for its pathogenicity and how to inhibit it.

The next few years will be an exciting, even nail-biting, time for those of us who have spent our scientific careers with at least a tacit assumption that genetic studies of actinomycetes might find applications. As described in Chapter 8, the decline in discovery of natural antibiotics does not imply that all the useful compounds have been found. It will be a huge and fascinating challenge to bring some of the untapped biosynthetic potential revealed by genome sequencing to light, a task that will succeed only through an increased effort to understand the overall biology of the organisms. This will stem from genomics in all its forms. It will be surprising and disappointing, too, if at least some of the promise of genetics and genetic engineering to generate medically useful unnatural natural products is not fulfilled.

Some academic scientists worry about the rise of genomics, remembering with nostalgia the apparently glorious early days of bacterial genetics, when a PhD student could work with a few Petri dishes to make interesting discoveries. They see genomics as "industrial" science, in which handles are mindlessly turned and results churned out. Scientists are becoming more interdependent and are feeling that, without access to genomic arrays, proteomic gels, and libraries of knockout mutants, they will be left out of a "race." As with all new developments, there is an upside and a downside.

It is no bad thing that communities of scientists with a common goal—such as gaining an integrated view of how a microorganism functions—should need to work together and share information as much as possible. As for the mindless element, I liken the current explosion of genomics to the post-Darwin burgeoning of evolutionary and taxonomic studies in the latter part of the 19th century. Biologists working with sharks or jellyfish, dandelions or moths, enthusiastically collected specimens from all over the world and described every detail of the anatomy and life cycles of their chosen animal or plant group in light of the unifying Darwinian idea that they had arisen by adaptive radiation from a recent common ancestor. Much of the work may not have been highly original, but it gave us a massive legacy of knowledge about the natural world that we would be lost without. Then, every so often, a new, creative insight broke through to a deeper level of understanding. The same will surely happen with genomics, where the collection and description of genes and their functions by the tools of the various "omics" outlined in Chapter 9 is equivalent to the 19th century comparative anatomy of plants and animals that depended on dissecting instruments and microscopes.

The current flurry of activity in *Streptomyces* genetics and genomics contrasts with the early years. New knowledge comes at a rate that could only be dreamed of when microbial genetics began with the isolation of mutants, the making of crosses, and attempts to squeeze a few drops of information about the arrangements of genes on the bacterial chromosome from incomplete data sets. Now, a graduate student can expect to make a significant or even dramatic advance in knowledge during a 3-year stint in the laboratory, and relate it to the broad field of biology through the wonders of comparative genomics. But that does not detract from the pleasure of looking back over a career in which intellectual effort and practical skills on the part of a succession of graduate students, postdoctoral scientists, and colleagues gradually laid the groundwork for what we know today about these beautiful, fascinating, and useful microbes. And, as I hope I have shown in this book, there is plenty still to do.

Notes and References

Chapter 1: Actinomycetes and Antibiotics

1. Waksman, S. A. (1954). *My life with the microbes*. London: Robert Hale.
2. Kupferberg, E. D. (2003). A field of great promise: soil bacteriology in America, 1900–1925. *Endeavour* 27, 16–21.
3. Waksman, S. A. (1927). *Principles of soil microbiology*. Baltimore: The Williams and Wilkins Company.
4. Harz, C. O. (1877–1878). *Actinomyces bovis,* ein neuer Schimmel in den Geweben des Rindes [*Actinomyces bovis*, a new fungus from cattle tissues]. *Jahresbericht der Kaiserlichen Central-Thierarznei-Schule in München,* 125–140.
5. Dr. Gunnar Odland, formerly consultant surgeon at Haukeland Hospital, Bergen; personal communication, 2003.
6. Hansen, G. A. (1874). Undersøgelser angående spedalskhedens årsager [Investigations on the causes of leprosy]. *Norsk Magazin for Lägervidenskaben* 9.
7. Bull, O. B. and Hansen, G. A. (1873). *The leprous diseases of the eye*. Christiania (Oslo): Albert Cammermeyer.
8. Drews, G. (1999). Ferdinand Cohn: a promoter of modern microbiology. *Nova Acta Leopoldina* 80, 13–43.
9. Cohn, F. (1875). Untersuchungen über Bakterien. [Investigations on bacteria] II. *Beiträge zur Biologie der Pflanzen* 1, 141–204.
10. Brock, T. D. (1999). *Robert Koch: a life in medicine and bacteriology*. Washington, DC: ASM Press.
11. Koch, R. (1882). Die Aetiologie der Tuberkulose [The cause of tuberculosis]. *Berliner klinische Wochenschrift* 19, 221–230. (English translation by Codell, K. C. [1987]. In

Contributions in Medical Studies 20, 83–96, New York: Greenwood Press.) This was Koch's first description of the tubercle bacillus; it was followed in 1884 by a comprehensive article under the same title in *Mitteilungen aus dem Kaiserlichem Gesundheitsamte* 2, 1–88. (English translation by Boyd, S. [1886]. In *Bacteria in relation to disease*, ed. Cheyne, W. W., 63–201. London: New Sydenham Society.)

12. Buchanan, R. E. (1916). Studies on the classification and nomenclature of the bacteria: VIII. The subgroups and genera of the Actinomycetales. *Journal of Bacteriology* 111, 403–406.

13. Henrici, A. (1930). *Molds, yeasts and actinomycetes*. New York: Wiley.

14. Waksman, S. A. and Henrici, A. T. (1943). The nomenclature and classification of the actinomycetes. *Journal of Bacteriology* 46, 337–341.

15. Lehmann, K. and Neumann, R. (1896). *Atlas und Grundriss der Bakteriologie* [Atlas and outline of bacteriology]. Munich: J. F. Lehmann.

16. Moberg, C. L. (2005). *René Dubos: friend of the good earth*. Washington DC: ASM Press.

17. Waksman was asked as early as 1941 to propose a term for the group of natural antimicrobial substances that were coming to light: see Pramer, D. (1988). Some choice words from Waksman. *The Scientist* 2, 17.

18. H. B. Woodruff, personal communication, January 2003.

19. Woodruff, H. B. (1981). A soil microbiologist's odyssey. *Annual Review of Microbiology* 35, 1–28.

20. *Of microbes and men*. Report of the Rutgers Research and Educational Foundation, June 1959.

21. Ryan, F. (1992). *Tuberculosis: the greatest story never told*. Bromsgrove: Swift Publishers.

22. Wainwright, M. (1991). Streptomycin: discovery and resultant controversy. *History and Philosophy of the Life Sciences* 13, 97–124; Kingston, W. (2004). Streptomycin, *Schatz v. Waksman*, and the balance of credit for discovery. *Journal of the History of Medicine and Allied Sciences* 59, 441–462.

Chapter 2: Antibiotic Discovery and Resistance

1. Kumazawa, J. and Yagisawa, M. (2002). The history of antibiotics: the Japanese story. *Journal of Infection and Chemotherapy* 8, 125–133.

2. Woodruff, H. B. (1999). Natural products from microorganisms: an odyssey revisited. *Actinomycetologica* 13, 58–67.

3. Luedemann, G. (1991). Free sprit of enquiry: the uncommon common man in research and discovery—the gentamicin story. *Actinomycetes* 2 (Suppl.1), new series, ed. Locci, R. Gorizia: The International Centre for Theoretical and Applied Ecology.

4. Benveniste, R. and Davies, J. (1973). Aminoglycoside antibiotic-inactivating enzymes in actinomycetes similar to those present in clinical isolates of antibiotic-resistant bacteria. *Proceedings of the National Academy of Sciences USA* 70, 2276–2280.

5. Joint Committee on the Use of Antibiotics in Animal Husbandry and Veterinary Medicine (1969). *[The Swann] Report*. London: Her Majesty's Stationery Office.

6. Amyes, S. G. B. (2001). *Magic bullets, lost horizons: the rise and fall of antibiotics*. London: Taylor and Francis.

7. Levy, S. B. (2002). *The antibiotic paradox: how the misuse of antibiotics destroys their curative powers*, 2nd ed. Cambridge, MA: Da Capo Press.

8. Wegener, H. C. (2003). Antibiotics in animal feed and their role in resistance development. *Current Opinion in Microbiology* 6, 439–445.

9. P. M. Hawkey, personal communication, April 2004.

Chapter 3: Microbial Sex

1. Davis, R. H. (2003). *The microbial models of molecular biology: from genes to genomes.* New York: Oxford University Press.

2. Kay, L. A. (1989). Selling pure science in wartime: the biochemical genetics of G. W. Beadle. *Journal of the History of Biology* 22, 73–101.

3. Waksman, S. A. (1050). *The actinomycetes: their nature, occurrence and importance.* Waltham, MA: Chronica Botanica Co.

4. Catcheside, D. G. (1951). *The genetics of micro-organisms*, London: Pitman.

5. Witkin, E. M. (2002). Chances and choices: Cold Spring Harbor 1944–1955. *Annual Review of Microbiology* 56, 1–15.

6. Glauert, A. (1986). The thirtieth birthday of Araldite. *Proceedings of the Royal Microscopical Society* 21, 375–377.

7. Cummins, C. S. and Harris, H. (1958). Studies on the cell-wall composition and taxonomy of Actinomycetales. *Journal of General Microbiology* 18, 173–189.

8. Woese, C. R. and Fox, G. E. (1977). Phylogenetic structure of the prokaryotic domain: the primary kingdoms. *Proceedings of the National Academy of Sciences USA* 74, 5088–5090.

9. Bignami, G. (2002). Origins and subsequent development of the Istituto Superiore di Sanità in Rome (Italy). *Annali di Igiene* 14 (Suppl. 1), 67–95.

10. Siddiqui, O. (2002). Guido Pontecorvo. *Biographical Memoirs of Fellows of the Royal Society* 48, 375–390.

11. Lisa Pontecorvo collection.

12. Brock, T. D. (1990). *The emergence of bacterial genetics.* Cold Spring Harbor, NY: Cold Spring Harbor Press.

13. Pontecorvo, G. (1976). Presidential address. In *Proceedings of the 2nd International Symposium on the Genetics of Industrial Microorganisms*, ed. MacDonald, K. D., 1–4. London: Academic Press.

14. Cairns, J. (1963). The chromosome of *Escherichia coli. Cold Spring Harbor Symposia on Quantitative Biology* 28, 43–46.

15. Stahl, F. W. (1967). Circular genetic maps. *Journal of Cellular Physiology* 70 (Suppl.1), 1–7. In this paper, Stahl discusses genetic map circularity in detail and makes the parenthetic comment that "nothing which can be said about circular maps is as important as what can be said within them." The significance of this comment must have escaped at least 99% of his readers, those who had not received from him a copy of one of his papers from the previous year in which he had printed "U. S. GET OUT OF VIET NAM" in the space inside a circular genetic map of a bacteriophage.

Chapter 4: Toward Gene Cloning

1. Olby, R. (1987). William Bateson's introduction of mendelism to England: a reassessment. *British Journal of the History of Science* 20, 399–420.

2. Vivian, A. (1971). Genetic control of fertility in *Streptomyces coelicolor* A3(2): plasmid involvement in the interconversion of UF and IF strains. *Journal of General Microbiology* 69, 353–364.

3. Fodor, K. and Alföldi, L. (1976). Fusion of protoplasts of *Bacillus megaterium*. *Proceedings of the National Academy of Sciences USA* 73, 47–50; Schaeffer, P., Cami, B. and Hotchkiss, R. D. (1976). Fusion of bacterial protoplasts. *Proceedings of the National Academy of Sciences USA* 73, 2151–2155.

4. Okanishi, M., Suzuki, K. and Umezawa, H. (1974). Formation and reversion of streptomycete protoplasts: cultural condition and morphological study. *Journal of General Microbiology* 80, 389–400.

5. Hales, A. (1977). A procedure for the fusion of cells in suspension by means of polyethylene glycol. *Somatic Cell Genetics* 3, 227–230.

6. Hopwood, D. A., Wright, H. M., Bibb, M. J. and Cohen, S. N. (1977). Genetic recombination through protoplast fusion in *Streptomyces*. *Nature* 268, 171–174.

7. Zhang, Y., Perry, K., Vinci, V. A. et al. (2002). Genome shuffling leads to rapid phenotypic improvement in bacteria. *Nature* 415, 644–646.

8. Cohen, S. N., Chang, A. C. Y. and Hsy, L. (1972). Nonchromosomal antibiotic resistance in bacteria: genetic transformation of *Escherichia coli* by R-factor DNA. *Proceedings of the National Academy of Sciences USA* 69, 2110–2114.

9. Bibb, M. J., Ward, J. M. and Hopwood, D. A. (1978). Transformation of plasmid DNA into *Streptomyces* at high frequency. *Nature* 274, 398–400.

10. Bibb, M. J., Schottel, J. L. and Cohen, S. N. (1980). A DNA cloning system for interspecific gene transfer in antibiotic-producing *Streptomyces*. *Nature* 284, 526–531; Thompson, C. J., Ward, J. M. and Hopwood, D. A. (1980). DNA cloning in *Streptomyces*: resistance genes from antibiotic-producing species. *Nature* 286, 525–527.

11. Sukhodolets, V. V. and Mkrtumian, N. M. (2003). Sos I. Alikhanian (1906–1985): founder of the Soviet School of Industrial Microbial Genetics. *Society for Industrial Microbiology News* 53, 214–218.

12. Lomovskaya, N. D., Mkrtumian, N. M. and Gostimskaya, N. L. (1970). Isolation and characteristics of *Streptomyces coelicolor* actinophage. [In Russian]. *Genetika* 6, 135–137.

13. Suarez, J. E. and Chater, K. F. (1980). DNA cloning in *Streptomyces*: a bifunctional replicon comprising pBR322 inserted into a *Streptomyces* phage. *Nature* 286, 527–529.

14. Mazodier, P., Petter, R. and Thompson, C. (1989). Intergeneric conjugation between *Escherichia coli* and *Streptomyces* species. *Journal of Bacteriology* 171, 3583–3585.

Chapter 5: From Chromosome Map to DNA Sequence

1. Smith, C. L. and Cantor, C. R. (1987). Purification, fragmentation and separation of large DNA molecules. *Methods in Enzymology* 155, 449–457; Carle, G. F. and Olson, M. V. (1987). Orthogonal-field-alternation gel electrophoresis. *Methods in Enzymology* 155, 468–482.

2. Kieser, H. M., Kieser, T. and Hopwood, D. A. (1992). A combined genetic and physical map of the *Streptomyces coelicolor* A3(2) chromosome. *Journal of Bacteriology* 174, 5496–5507.

3. Lin, Y. S., Kieser, H. M., Hopwood, D. A. and Chen, C. W. (1993). The chromosomal DNA of *Streptomyces lividans* 66 is linear. *Molecular Microbiology* 10, 923–933.

4. Judson, O. P. and Haydon, D. (1999). The genetic code: what is it good for? An analysis of the effects of selection pressures on genetic codes. *Journal of Molecular Evolution* 49, 539–550.

5. Crick, F. H., Barnett, L, Brenner, S. and Watts-Tobin, R. J. (1961). General nature of the genetic code for proteins. *Nature* 192, 1227–1232.

6. *The Arabian Nights: Tales from a Thousand and One Nights*, trans. R. Burton. New York: Random House Modern Library, paperback edition 2004.

7. Bentley, S, D., Chater K. R., Cerdeno-Tarraga, A. M. et al. (2002). Complete genome sequence of the model actinomycete *Streptomyces coelicolor* A3(2). *Nature* 417, 141–147.

8. Blattner, F. R., Plunkett, G., Block, C. A. et al. (1997). The complete genome sequence of *Escherichia coli* K-12. *Science* 277, 1453–1474.

9. Cole, S. T., Brosch, R., Parkhill, J. et al. (1998). Deciphering the biology of *Mycobacterium tuberculosis* from the complete genome sequence. *Nature* 393, 537–544.

10. Nakamura, Y., Oda, T., Matsuda, H. and Gojobori, T. (2004). Biased biological functions of horizontally transferred genes in prokaryotic genomes. *Nature Genetics* 36, 760–766; Jain, R., Rivera, M. C., Moore, J. E. and Lake, J. A. (2002). Horizontal gene transfer in microbial genome evolution. *Theoretical Population Biology* 61, 489–495; Dutta, C. and Pan, A (2002). Horizontal gene transfer and bacterial diversity. *Journal of Bioscience* 27 (Suppl.1), 27–33.

11. Ikeda, H., Ishikawa, J., Hanamoto, A. et al. (2003). Complete genome sequence and comparative analysis of the industrial microorganism *Streptomyces avermitilis*. *Nature Biotechnology* 21, 526–531.

Chapter 6: Bacteria That Develop

1. Grund, A. D. and Ensign, J. C. (1985). Properties of the germination inhibitor of *Streptomyces viridochromogenes* spores. *Journal of General Microbiology* 131, 833–847.

2. Claessen, D., Rink, R., de Jong, W. et al. (2003). A novel class of secreted hydrophobic proteins involved in aerial mycelium formation in *Streptomyces coelicolor* by forming amyloid-like fibrils. *Genes and Development* 17, 1714–1726; Elliot, M., Karoonuthaisiri, N., Huang, J. et al. (2003). The chaplins: a family of hydrophobic cell-surface proteins involved in aerial mycelium development in *Streptomyces coelicolor*. *Genes and Development* 17, 1727–1740.

3. Claessen, D., Stokroos, I., Deelstra, H. J. et al. (2004). The formation of the rodlet layer of streptomycetes is the result of the interplay between rodlins and chaplins. *Molecular Microbiology* 53, 433–443.

4. van Keulen, G., Hopwood, D. A., Dijkhuizen, L. and Sawers, R. G. (2005). Gas vesicles: old buoys in novel habitats. *Trends in Microbiology* 13, 350–354.

5. Krištůfek, V., Hallmann, M., Wesheide, W. and Schrempf, H. (1995). Selection of various *Streptomyces* species by *Enchytraeus crypticus* (Oligochaeta). *Pedobiologia* 39, 547–554.

6. Currie, C. R., Scott, J. A., Summerbell, R. C. and Malloch, D (1999). Fungus-growing ants use antibiotic-producing bacteria to control garden parasites. *Nature* 398, 701–704.

7. AgBio, Inc. Mycostop: A biological fungicide developed from a naturally occurring biocontrol microbe. Available at: http://www.agbio-inc.com/Mycostop.htm (accessed April 15, 2006).

8. Tokala, R. K., Strap, J. L., Jung, C. M. et al. (2002). Novel plant-microbe rhizosphere interaction involving *Streptomyces lydicus* WYEC108 and the pea plant (*Pisum sativum*). *Applied and Environmental Microbiology* 68, 2161–2171.

9. Castillo, U. F., Strobel, G. A., Ford, E. J. et al. (2002). Munumbicins, wide-spectrum antibiotics produced by *Streptomyces* NRRL 30562, endophyte on *Kennedia nigriscans*. *Microbiology* 148, 2675–2685.

10. McCormick, J. R., Su, E. P., Driks, A. and Losick, R. (1994). Growth and viability of *Streptomyces coelicolor* mutant for the cell division gene *FtsZ*. *Molecular Microbiology* 14, 243–254.

11. Jakimowicz, D., Majkadagger, J., Konopa, G. et al. (2000). Architecture of the *Streptomyces lividans* DnaA protein-replication origin complexes. *Journal of Molecular Biology* 298, 351–364.

12. Bao, K. and Cohen, S. N. (2003). Recruitment of terminal protein to the ends of *Streptomyces* linear plasmids and chromosomes by a novel telomere-binding protein essential for linear DNA replication. *Genes and Development* 17, 774–785.

13. Jakimowicz, D., Chater, K. and Zakrzewska-Czerwinska, J. (2002). The ParB protein of *Streptomyces coelicolor* A3(2) recognizes a cluster of *parS* sequences within the origin-proximal region of the linear chromosome. *Molecular Microbiology* 45, 1365–1377; Jakimowicz, D., Gust, B., Zakrzewska-Czerwinska, J. and Chater, K. F. (2005). Developmental-stage-specific assembly of ParB complexes in *Streptomyces coelicolor* hyphae. *Journal of Bacteriology* 187, 3572–3580.

14. Reuther, J., Gekeler C., Tiffert, Y. et al. (2006). Unique conjugation mechanism in mycelial streptomycetes: a DNA binding, ATPase translocates DNA at the hyphal tip. *Molecular Microbiology* 61, 436–446.

15. Errington, J., Bath, J. and Wu, L. J. (2001). DNA transport in bacteria. *Nature Reviews in Molecular and Cell Biology* 2, 538–545.

16. Possoz, C., Ribard C., Gagnat, J. et al. (2001). The integrative element pSAM2 from *Streptomyces*: kinetics and mode of conjugal transfer. *Molecular Microbiology* 42, 159–166.

17. Yamasaki, M. and Kinashi, H. (2004). Two chimeric chromosomes of *Streptomyces coelicolor* A3(2) generated by single crossover of the wild-type chromosome and linear plasmid SCP1. *Journal of Bacteriology* 186, 6553–6559.

18. Huang, C. H., Lin, Y. S., Yang, Y. L. et al. (1998). The telomeres of *Streptomyces* chromosomes contain conserved palindromic sequences with potential to form complex secondary structures. *Molecular Microbiology* 28, 905–916.

19. Yang, M. C. and Losick, R. (2001). Cytological evidence for association of the ends of the linear chromosome in *Streptomyces coelicolor*. *Journal of Bacteriology* 183, 5180–5186.

20. Hopwood, D. A. (2006). Soil to genomics: the *Streptomyces* chromosome. *Annual Review of Genetics* 40 (in press).

21. Kodani, S., Hudson, M. E., Durrant, M. C. et al. (2004). The SapB morphogen is a lantibiotic-like peptide derived from the product of the developmental gene *ramS* in *Streptomyces coelicolor*. *Proceedings of the National Academy of Sciences USA* 101, 11448–11453.

22. Wösten, H. J. & Willey, J. M. (2000). Surface-active proteins enable microbial aerial hyphae to grow into the air. *Microbiology* 146, 767–773.

23. Miguelez, E. M., Hardisson, C. and Manzanal, M. B. (1999). Hyphal death during colony development in *Streptomyces antibioticus*: morphological evidence for the existence of a process of cell deletion in a multicellular prokaryote. *Journal of Cell Biology* 145, 515–525.

24. Kim, I. S. and Lee, K. J. (1995). Physiological roles of leupeptin and extracellular proteases in mycelium development of *Streptomyces exfoliatus* SMF13. *Microbiology* 141, 1017–1025.

25. Jacob, F. and Monod, J. (1961). Genetic regulatory mechanisms in the synthesis of proteins. *Journal of Molecular Biology* 3, 318–356.

26. Englesberg, E., Irr, J., Power, J. and Lee, N. (1965). Positive control of enzyme synthesis by gene C in the L-arabinose system. *Journal of Bacteriology* 90, 946–957.

27. Westpheling, J., Ranes, M. and Losick, R. (1985). RNA polymerase heterogeneity in *Streptomyces coelicolor*. *Nature* 313, 22–27.

28. Lawlor, E. J., Baylis, H. A. and Chater, K. F. (1987). Pleiotropic morphological and antibiotic deficiencies result from mutations in a gene encoding a tRNA-like product in *Streptomyces coelicolor* A3(2). *Genes and Development* 1, 1305–1310.

29. Takano, E., Tao, M., Long, F. et al. (2003). A rare leucine codon in *adpA* is implicated in the morphological defect of *bldA* mutants of *Streptomyces coelicolor*. *Molecular Microbiology* 50, 475–486; Nguyen, K. T., Tenor, J., Stettler, H. et al. (2003). Colonial differentiation in *Streptomyces coelicolor* depends on translation of a specific codon within the *adpA* gene. *Journal of Bacteriology* 185, 7291–7296.

30. Chater, K. F., Bruton, C. J., Plaskitt, K. A. et al. (1989). The developmental fate of *S. coelicolor* hyphae depends upon a gene product homologous with the motility sigma factor of *B. subtilis*. *Cell* 59, 133–143.

31. Losick, R. and Pero, J. (1981). Cascades of sigma factors. *Cell* 25, 582–584.

32. Chater, K. F. (1998). Taking a genetic scalpel to the *Streptomyces* colony. *Microbiology* 144, 1465–1478.

33. Bruton, C. J., Plaskitt, K. A. and Chater, K. F. (1995). Tissue-specific glycogen branching isoenzymes in a multicellular prokaryote, *Streptomyces coelicolor* A3(2). *Molecular Microbiology* 18, 89–99.

Chapter 7: The Switch to Antibiotic Production

1. Locher, K. P. (2004). Structure and mechanism of ABC transporters. *Current Opinion in Structural Biology* 14, 424–431.

2. Doyle, D. A, Morais Cabral, J., Pfuetzner, R. A. et al. (1998). The structure of the potassium channel: molecular basis of K^+ conduction and selectivity. *Science* 280, 69–77.

3. Virolle, M. J. and Bibb, M. J. (1988). Cloning, characterization and regulation of an alpha-amylase gene from *Streptomyces limosus*. *Molecular Microbiology* 2, 197–208.

4. Ni, X. and Westpheling, W. (1997). Direct repeats in the *Streptomyces* chitinase-63 promoter direct both glucose repression and chitin induction. *Proceedings of the National Academy of Sciences USA* 94, 13116–13121.

5. Walter, S. and Schrempf, H. (2003). Oligomerization, membrane anchoring, and cel-

lulose-binding characteristics of AbpS, a receptor-like *Streptomyces* protein. *Journal of Biological Chemistry* 278, 26639–26647.

6. Challis, G. L. and Hopwood, D. A. (2003). Synergy and contingency as driving forces for the evolution of multiple secondary metabolite production by *Streptomyces* species. *Proceedings of the National Academy of Sciences USA* 100 (Suppl. 2), 14555–14561.

7. Hong, H. J., Hutching, M. I., Nu, J. M. et al. (2004). Characterization of an inducible vancomycin resistance system in *Streptomyces coelicolor* reveals a novel gene (*vanK*) required for drug resistance. *Molecular Microbiology* 52, 1107–1121.

8. Lonetto, M. A., Brown, K. L., Rudd, K. E. and Buttner, M. J. (1994). Analysis of the *Streptomyces coelicolor sigE* gene reveals the existence of a subfamily of eubacterial RNA polymerase sigma factors involved in the regulation of extracytoplasmic functions. *Proceedings of the National Academy of Sciences USA* 91, 7573–7577.

9. Paget, M. S., Molle, V., Cohen, G. et al. (2001). Defining the disulphide stress response in *Streptomyces coelicolor* A3(2): identification of the sigmaR regulon. *Molecular Microbiology* 42, 1007–1020.

10. Chater, K. F. and Merrick, M. J. (1979). Streptomycetes. In *Developmental Biology of Prokaryotes*, ed. Parrish, J. H., 93–114. Oxford: Blackwell.

11. Arias, P., Fernández-Moreno, M. A. and Malpartida, F. (1999). Characterization of the pathway-specific positive transcriptional regulator for actinorhodin biosynthesis in *Streptomyces coelicolor* A3(2) as a DNA-binding protein. *Journal of Bacteriology* 181, 6958–6968.

12. Kirby, R., Wright, L. F. and Hopwood, D. A. (1975). Plasmid-determined antibiotic synthesis and resistance in *Streptomyces coelicolor*. *Nature* 254, 265–267.

13. Suwa, M., Sugino, H., Sasaoka, A. et al. (2000). Identification of two polyketide synthase gene clusters on the linear plasmid pSLA2–L in *Streptomyces rochei*. *Gene* 246, 123–131.

14. Wietzorrek, A. and Bibb, M. (1997). A novel family of proteins that regulates antibiotic production in streptomycetes appears to contain an OmpR-like DNA-binding fold. *Molecular Microbiology* 25, 1181–1184.

15. Piepersberg, W. (1995). Streptomycin and related aminoglycosides. In *Genetics and biochemistry of antibiotic production*, ed. Vining, L. C. and Stuttard, C., 531–570. Boston: Butterworth-Heinemann.

16. Fernández-Moreno, M. A., Caballero, J. L., Hopwood, D. A. and Malpartida, F. (1991). The *act* cluster contains regulatory and antibiotic export genes, direct targets for translational control by the *bldA* tRNA gene of *Streptomyces*. *Cell* 66, 769–780.

17. Horinouchi, S. (2002). A microbial hormone, A-factor, is a master switch for morphological differentiation and secondary metabolism in *Streptomyces griseus*. *Frontiers in Bioscience* 7, d2045–d2057.

18. Haygood, M. G. (1993). Light organ symbioses in fishes. *Critical Reviews in Microbiology* 19, 191–216.

19. Takano, E. (2006). Gamma-butyrolactones: *Streptomyces* signalling molecules regulating antibiotic production and differentiation. *Current Opinion in Microbiology* 9, 1–8.

Chapter 8: Unnatural Natural Products

1. Hopwood, D. A., Malpartida, F., Kieser, H. M. et al. (1985). Production of "hybrid" antibiotics by genetic engineering. *Nature* 314, 642–644.

2. Wakil, S. J. (1989). Fatty acid synthase, a proficient multifunctional enzyme. *Biochemistry* 28, 4523–4530.

3. Sherman, D. H., Malpartida, F., Bibb, M. J. et al. (1989). Structure and deduced function of the granaticin-producing polyketide synthase gene cluster of *Streptomyces violaceoruber* Tü22. *EMBO Journal* 8, 2717–2725; Bibb, M. J., Biro, S., Motamedi, H. et al. (1989). Analysis of the nucleotide sequence of the *Streptomyces glaucescens tcmI* genes provides key information about the enzymology of polyketide antibiotic biosynthesis. *EMBO Journal* 8, 2727–2736; Fernández-Moreno, M. A., Martinez, E. Boto, L. et al. (1992). Nucleotide sequence and deduced functions of a set of co-transcribed genes of *Streptomyces coelicolor* A3(2) including the polyketide synthase for the antibiotic actinorhodin. *Journal of Biological Chemistry* 267, 19278–19290.

4. Sherman, D. H., Kim, E. S., Bibb, M. J. and Hopwood, D. A. (1992). Functional replacement of genes for individual polyketide synthase components in *Streptomyces coelicolor* A3(2) by heterologous genes from a different polyketide pathway. *Journal of Bacteriology* 174, 6184–6190.

5. Bartel, P. L., Zhu, C. B., Lampel, J. S. et al. (1990). Biosynthesis of anthraquinones by interspecies cloning of actinorhodin biosynthesis genes in streptomycetes: clarification of actinorhodin gene functions. *Journal of Bacteriology* 172, 4816–4826.

6. McDaniel, R., Ebert-Khosla, S., Hopwood, D. A. and Khosla, C. (1995). Rational design of aromatic polyketide natural products by recombinant assembly of enzymatic subunits. *Nature* 375, 549–554.

7. Cortes, J., Haydoc, S. F., Roberts, G. A. et al. (1990). An unusually large multifunctional polypeptide in the erythromycin-producing polyketide synthase of *Saccharopolyspora erythraea*. *Nature* 348, 176–178; Donadio, S., Staver, M. J., McAlpine, J. B. et al. (1991). Modular organization of genes required for complex polyketide biosynthesis. *Science* 252, 675–679.

8. Kao, C. M., Katz, L. and Khosla, C. (1994). Engineered biosynthesis of a complete macrolactone in a heterologous host. *Science* 265, 509–512.

9. Haydock, S. F., Aparicio, J. F., Molnar, I. et al. (1995). Divergent sequence motifs correlated with the substrate specificity of (methyl)malonyl-CoA:acyl carrier protein transacylase domains in modular polyketide synthases. *FEBS Letters* 374, 246–248.

10. Revill, W. P., Voda, J., Reeves, C. R. et al. (2002). Genetically engineered analogs of ascomycin for nerve regeneration. *Journal of Pharmacology and Experimental Therapeutics* 302, 1278–1285.

11. Abbanat, D., Webb, G., Foleno, B. et al. (2005). *In vitro* activities of novel 2-fluoronaphthyridine-containing ketolides. *Antimicrobial Agents and Chemotherapy* 49, 309–315.

12. Zazopoulos, E., Huang, K., Staffa, A. et al. (2003). A genomics-guided approach for discovering and expressing cryptic metabolic pathways. *Nature Biotechnology* 21, 187–190.

13. Watve, M. G., Tickoo, R., Jog, M. M. and Bhole, B. D. (2001). How many antibiotics are produced by the genus *Streptomyces*? *Archives of Microbiology* 176, 386–390.

14. Kelner, A. (1949). Studies on the genetics of antibiotic formation: the induction of antibiotic-forming mutants in actinomycetes. *Journal of Bacteriology* 57, 73–92.

15. Sezonov, G., Blanc, V., Bamas-Jacques, N. et al. (1997). Complete conversion of antibiotic precursor to pristinamycin IIA by overexpression of *Streptomyces pristinaespiralis* biosynthetic genes. *Nature Biotechnology* 15, 349–353.

16. Stutzman-Engwall, K., Conlon, S., Fedechko, R. et al. (2005). Semi-synthetic DNA shuffling of *aveC* leads to improved industrial scale production of doramectin by *Streptomyces avermitilis*. *Metabolic Engineering* 7, 27–37.

Chapter 9: Functional Genomics

1. Fleischmann, R. D., Adams, M. D., White, O. et al. (1995). Whole-genome random sequencing and assembly of *Haemophilus influenzae* Rd. *Science* 269, 496–512.

2. Mewes, H. W., Albermann, K., Bahr M. et al. (1997). Overview of the yeast genome. *Nature* 387 (Suppl.), 7–9.

3. Genomes On Line Database. Available at: http://www.genomesonline.org (accessed April 15, 2006).

4. Mullis, K. B. (1990). The unusual origin of the polymerase chain reaction. *Scientific American* 262, 56–65.

5. Brown, P. O. and Botstein, D. (1999). Exploring the new world of the genome with DNA microarrays. *Nature Genetics* 21 (Suppl.), 33–37.

6. Huang, J., Lih, C. J., Pan, K. H. and Cohen. S. N. (2001). Global analysis of growth phase responsive gene expression and regulation of antibiotic biosynthetic pathways in *Streptomyces coelicolor* using DNA microarrays. *Genes and Development* 15, 3183–3192.

7. O'Farrell, P. H. (1975). High resolution two-dimensional electrophoresis of proteins. *Journal of Biological Chemistry* 250, 4007–49021.

8. Hesketh, A. R., Chandra, G., Shaw A. D. et al. (2002). Primary and secondary metabolism, and post-translational protein modifications, as portrayed by proteomic analysis of *Streptomyces coelicolor*. *Molecular Microbiology* 46, 917–932.

9. McClintock, B. (1950). The origin and behavior of mutable loci in maize. *Proceedings of the National Academy of Sciences USA* 36, 344–355.

10. Jordan, E., Saedler, H. and Starlinger, P. (1968). O° and strong polar mutations in the *gal* operon are insertions. *Molecular and General Genetics* 102, 353–363; Shapiro, J. A. (1969). Mutations caused by the insertion of genetic material into the galactose operon of *Escherichia coli*. *Journal of Molecular Biology* 40, 93–105.

11. Watanabe, T. (1963). Infective heredity of multiple drug resistance in bacteria. *Bacteriological Reviews* 27, 87–115.

12. Lamb, D. C., Fowler, K., Kieser, T. et al. (2002). Sterol 14alpha-demethylase activity in *Streptomyces coelicolor* A3(2) is associated with an unusual member of the CYP51 gene family. *Biochemical Journal* 364, 555–562.

13. Gust, B., Chandra, G., Jakimowicz, D. et al. (2004). Lambda red-mediated genetic manipulation of antibiotic-producing *Streptomyces*. *Advances in Applied Microbiology* 54, 107–128.

14. Datsenko, K. A. and Wanner, B. L. (2000). One-step inactivation of chromosomal genes in *Escherichia coli* K-12 using PCR products. *Proceedings of the National Academy of Sciences USA* 97, 6640–6645.

Chapter 10: Genomics Against Tuberculosis and Leprosy

1. Morel, C. M. (2000). Reaching maturity: 25 years of the TDR. *Parasitology Today* 16, 522–528; Anonymous. (1996). Four TDR diseases can be "eliminated." 2. Leprosy: from elimination to eradication? *TDR News* 49, 3.

2. W. R. Jacobs, personal communication, February 2005.

3. Jacobs, W. R., Tuckman, M. and Bloom, B. R. (1987). Introduction of foreign DNA into mycobacteria using a shuttle plasmid. *Nature* 327, 532–535.

4. Rawcliffe, C. (2004). Sickness and health. In *Medieval Norwich*, ed. Rawcliffe, C. and Wilson, R., 301–326. London and New York: London & Hambledon.

5. Daniel, T. M., Bates, J. H. and Downes, K. A. (1994). History of tuberculosis. In *Tuberculosis: pathogenesis and control*, ed. Bloom, B. R., 13–24. Washington, DC: ASM Press.

6. Ryan, F. (1992). *Tuberculosis: the greatest story never told*. Bromsgrove: Swift Publishers.

7. Atkinson, P., Taylor, H., Sharland, M. and Maguire, H. (2002). Resurgence of paediatric tuberculosis in London. *Archives of Disease in Children* 86, 264–265.

8. J. Colston, personal communication, May 2002.

9. Webb, V. and Davies, J. (1999). Antibiotics and antibiotic resistance in mycobacteria. In *Mycobacteria: molecular biology and virulence*, eds. Ratledge, C., and Dale, J., 187–305. Oxford: Blackwell Science.

10. Dye, C., Williams, B. G., Espinal, M. A. and Raviglione, M. C. (2002). Erasing the world's slow stain: strategies to beat multidrug-resistant tuberculosis. *Science* 295, 2042–2046.

11. Calmette, A. and Guérin, C. (1920). Nouvelles recherches expérimentales sur la vaccination des bovidés contre la tuberculose. *Annales de l'Institut Pasteur* 34, 553–560.

12. Lowrie, D. B. (1999). Vaccines. In *Mycobacteria: molecular biology and virulence* ed. Ratledge, C. and Dale, J., 335–349. Oxford: Blackwell Science.

13. Andersen, P. and Doherty, T. M. (2005). The success and failure of BCG: implications for a novel tuberculosis vaccine. *Nature Reviews Microbiology* 3, 565–662.

14. Cole, S. T., Brosch, R., Parkhill, J. et al. (1998). Deciphering the biology of *Mycobacterium tuberculosis* from the complete genome sequence. *Nature* 393, 537–544; Cole, S. T., Eiglmeier, K., Parkhill, J. et al. (2001). Massive gene decay in the leprosy bacillus. *Nature* 409, 1007–1011.

15. Sreevatsan, S., Pan, X., Stockbauer, K. E. et al. (1997). Restricted structural gene polymorphism in the *Mycobacterium tuberculosis* complex indicates evolutionarily recent global dissemination. *Proceedings of the National Academy of Sciences USA* 94, 9869–9874.

16. Brosch, R., Gordon, S. V., Marmiesse, M. et al. (2002). A new evolutionary scenario for the *Mycobacterium tuberculosis* complex. *Proceedings of the National Academy of Sciences USA* 99, 3684–3689.

17. Behr, M. A., Wilson, M. A., Gill, W. P. et al. (1999). Comparative genomics of BCG vaccines by whole-genome DNA microarray. *Science* 284, 1520–1523.

18. Hochhut, B., Dobrindt, U. and Hacker, J. (2005). Pathogenicity islands and their role in bacterial virulence and survival. *Contributions to Microbiology* 12, 234–254.

19. Hacker, J., Blum-Oehler, G., Mühldorfer, I. and Tschäpe (1997). Pathogenicity islands of virulent bacteria: structure, function and impact on microbial evolution. *Molecular Microbiology* 23, 1089–1097.

20. Hayashi, T., Makimo, K., Ohnishi, M. et al. (2001). Complete genome sequence of enterohemorrhagic *Escherichia coli* O157:H7 and genomic comparison with a laboratory strain K-12. *DNA Research* 8, 11–22.

21. Kers, J. A., Cameron, K. D., Joshi, M. V. et al. (2005). A large, mobile pathogenicity island confers plant pathogenicity on *Streptomyces* species. *Molecular Microbiology* 55, 1025–1033.

22. Larsen, M. H., Vilcheze, C., Kremer, L. et al. (2002). Overexpression of *inhA*, but not *kasA*, confers resistance to isoniazid and ethionamide in *Mycobacterium smegmatis*, *M. bovis* BCG and *M. tuberculosis*. *Molecular Microbiology* 46, 453–466.

23. Cole, S. T., Brosch, R., Parkhill, J. et al. (1998). Deciphering the biology of *Mycobacterium tuberculosis* from the complete genome sequence. *Nature* 393, 537–544; Cole, S. T., Eiglmeier, K., Parkhill, J. et al. (2001). Massive gene decay in the leprosy bacillus. *Nature* 409, 1007–1011.

24. Goyal, M., Shaw, R. J., Banerjee, D. K. et al. (1997). Rapid detection of multidrug-resistant tuberculosis. *European Respiratory Journal* 10, 1120–1124.

25. Global Alliance for TB Drug Development. Available at: http://www.who.int/tdr/diseases/tb/tballiance.htm (accessed April 15, 2006).

26. Wilson, M., DeRisi, J., Kristensen, H. H. et al. (1999). Exploring drug-induced alterations in gene expression in *Mycobacterium tuberculosis* by microarray hybridization. *Proceedings of the National Academy of Sciences USA* 96, 12833–12838.

27. Dubnau, E. and Smith, I. (2003). *Mycobacterium tuberculosis* gene expression in macrophages. *Microbes and Infection* 5, 629–637.

28. Fisher, M. A., Plikaytis, B. B. and Shinnick, T. M. (2002). Microarray analysis of the *Mycobacterium tuberculosis* transcriptional response to the acidic conditions found in phagosomes. *Journal of Bacteriology* 184, 4025–4032.

29. McKinney, J. D., Honer zu Bentrup, K., Muñoz-Elias, E. T. et al. (2000). Persistence of *Mycobacterium tuberculosis* in macrophages and mice requires the glyoxylate shunt enzyme isocitrate lyase. *Nature* 406, 683–685.

30. Soliveri, J. A., Gomez, J., Bishai, W. R. and Chater, K. F. (2000). Multiple paralogous genes related to the *Streptomyces coelicolor* developmental regulatory gene *whiB* are present in *Streptomyces* and other actinomycetes. *Microbiology* 146, 333–343.

31. Morris, R. P., Nguyen, L., Gatfield, J. et al. (2005). Ancestral antibiotic resistance in *Mycobacterium tuberculosis*. *Proceedings of the National Academy of Sciences USA* 102, 12200–12205.

32. Andries, K., Verhasselt, P., Guillemont, J. et al. (2005). A diarylquinoline drug active on the ATP synthase of *Mycobacterium tuberculosis*. *Science* 307, 223–227.

Glossary

SMALL CAPS are used for terms within definitions that are defined elsewhere in the Glossary.

Acyl carrier protein (ACP) The subunit of a FATTY ACID SYNTHASE or POLYKETIDE SYNTHASE that carries the growing carbon chain during its biosynthesis.
Adenine (A) One of the four BASES making up DNA and RNA; it pairs with thymine (T) in the double helix of DNA.
Amino acids The 20 kinds of building units of PROTEINS.
Auxotroph A mutant microorganism with a block in the biosynthetic pathway for a nutrient such as an amino acid or vitamin, which therefore must be added to the growth medium (contrast PROTOTROPH).
Bacteriophage (phage) A bacterial virus.
Bases The subunits of DNA and RNA whose order along the molecules determines their specificity to encode a PROTEIN sequence via the rules of the genetic code.
Chromosome The structure that carries the genes. In bacteria, it is a DNA molecule associated with only a few proteins, whereas in EUKARYOTES the chromosomes consist of a DNA molecule combined with several proteins, making a complex called chromatin.
Codon A triplet of bases on the mRNA corresponding to a particular AMINO ACID by the rules of the genetic code.
Coenzyme A (CoA) A molecule that activates the organic acids used to build FATTY ACIDS and POLYKETIDES; thus we speak of acetyl-CoA, malonyl-CoA, and so on.
Condensation The joining together of two organic compounds with loss of two hydrogen atoms and an oxygen atom as a molecule of water (H_2O).

Conjugation The process of mating in bacteria such as *Escherichia coli* or *Streptomyces*, promoted by PLASMIDS.

Cosmid A special kind of PLASMID cloning VECTOR that can be converted *in vitro* into a BACTERIOPHAGE, thereby allowing the easy introduction of foreign DNA into *Escherichia coli* by the natural process of phage infection.

Crossing-over The process in which DNA molecules of similar sequence exchange corresponding regions, resulting in genetic recombination. It occurs regularly between CHROMOSOMES at meiosis in EUKARYOTES and after DNA transfer by TRANSFORMATION, TRANSDUCTION, or CONJUGATION in bacteria.

Cytoplasm The contents of a cell other than the nucleus (in a EUKARYOTE) or the CHROMOSOME (in a PROKARYOTE).

Cytosine (C) One of the four BASES making up DNA and RNA; it pairs with guanine (G) in the double helix of DNA.

Dehydration Removal from an organic compound of two hydrogen atoms and an oxygen atom as a water molecule (H_2O) by an enzyme called a DEHYDRATASE. The water often derives from a hydroxyl group (–OH) attached to one carbon atom and a hydrogen (H) from an adjacent carbon atom.

DNA polymerase The enzyme that replicates DNA.

Eukaryotes Organisms, such as fungi, plants, and animals, that have a true nucleus consisting of a nuclear envelope containing CHROMOSOMES (contrast bacteria and archaea, which are PROKARYOTES).

Fatty acids A class of hydrocarbons consisting of carbon chains carrying a full or nearly full complement of hydrogen atoms; components of LIPIDS.

Fatty acid synthase (FAS) The enzyme that builds FATTY ACIDS from organic acids such as acetate and malonate while activated by attachment to COENZYME A. The FAS is multifunctional, consisting of acyl transferase, ketosynthase, ACYL CARRIER PROTEIN, keto-REDUCTASE, DEHYDRATASE, and enoyl-REDUCTASE functions.

Gene A unit of inheritance, consisting of a stretch of DNA that usually encodes a PROTEIN; exceptions are the genes that produce ribosomal RNA, the RNA molecules that form part of the structure of the RIBOSOMES, and the TRANSFER RNAs (tRNAs).

Genome The DNA that carries the full set of genes in an organism. In bacteria, with a single chromosome, the terms CHROMOSOME and GENOME are often used synonymously.

Genomics Study of the complete set of genes in an organism.

Gram-negative/positive The two major groups of bacteria, distinguished by the presence/absence of an extra membrane outside the cell wall.

Guanine (G) One of the four BASES making up DNA and RNA; it pairs with cytosine (C) in the double helix of DNA.

Hydrophilic Having an affinity for water (contrast HYDROPHOBIC, having an affinity for LIPIDS).

Hydrophobic Having an affinity for lipids (contrast HYDROPHILIC and see also LIPOPHILIC)

Hypha (*plural* **hyphae**) An elongated, branching cell characteristic of actinomycetes and fungi.

Intron A stretch of DNA within the coding sequence of a eukaryotic gene that is removed (spliced out) from the RNA transcript to give the mature MESSENGER RNA (mRNA).

Keto-group The chemical grouping C=O.

Lipids The class of organic molecules that includes fats and oils.
Lipophilic Having an affinity for LIPIDS.
Meiosis and **mitosis** The two types of nuclear division in EUKARYOTES. Meiosis produces progeny in which the diploid number of CHROMOSOMES is halved to the haploid number, whereas mitosis yields nuclei identical to the starting nucleus.
Messenger RNA (mRNA) The molecules of RNA produced from a DNA template by the RNA POLYMERASE in the process of TRANSCRIPTION of the genes. The mRNA goes on to be translated into PROTEIN on the RIBOSOME.
Mutagen A chemical or a physical agent such as X-rays or light ultraviolet light that induces MUTATIONS.
Mutation The process in which a daughter DNA molecule comes to differ from the parent by substitution, addition, or deletion of one or a small number of BASES.
Mycelium The mass of HYPHAE that makes up an actinomycete or fungus colony.
Nucleic acids The class of compounds that includes DNA and RNA.
Nucleotide A building unit of NUCLEIC ACIDS, consisting of a base (A, C, G, T, or U) joined to a sugar (deoxyribose for DNA, ribose for RNA) and a triphosphate group.
Operator A short DNA sequence upstream of a gene, to which a transcription factor binds, thereby facilitating or blocking access of the RNA POLYMERASE to the PROMOTER and switching a gene on or off.
Operon A group of two or more adjacent genes in a bacterium that are cotranscribed into the same mRNA molecule.
Pulsed field gel electrophoresis (PFGE) A technique to separate very large DNA molecules according to size on an agarose gel.
Plasmid A DNA molecule, separate from the main bacterial CHROMOSOME, which carries genes that are not essential under all conditions but confer an advantage to the host under certain circumstances; sex (or "fertility") plasmids promote CONJUGATION in bacteria, including *Escherichia coli* and *Streptomyces*.
Polyketides A huge class of natural products, including many antibiotics and anti-cancer agents, made by the assembly of organic acids such as acetic acid, malonic acid, and methylmalonic acid, while activated by attachment to COENZYME A, into long carbon chains.
Polyketide synthase (PKS) An enzyme related to a FATTY ACID SYNTHASE that builds a POLYKETIDE.
Primer A stretch of single-stranded RNA or DNA, complementary to part of a DNA "template" molecule, that allows DNA POLYMERASE to begin replicating the template DNA.
Probe A sequence of DNA or RNA, labeled either radioactively or with a colored dye, used to recognize a complementary sequence.
Prokaryotes Bacteria and archaea, distinguished from EUKARYOTES by having no nuclear envelope separating the CHROMOSOME from the rest of the cell.
Promoter The DNA sequences upstream of genes to which the RNA POLYMERASE binds via its SIGMA FACTOR subunit to initiate TRANSCRIPTION.
Proteins Molecules made up of a chain of AMINO ACIDS whose sequence determines the shape and reactivity of the protein, and therefore the role it plays; most proteins are ENZYMES that catalyze chemical reactions, but some make up the structure of cells and organisms.

Proteome The complete set of proteins in an organism.

Protoplast A cell of a gram-positive bacterium such as *Streptomyces*, or a plant, from which the wall has been removed.

Prototroph A microorganism that is able to synthesize all the complex nutrients that it needs, such as amino acids and vitamins, which therefore do not need to be added to the growth medium (contrast AUXOTROPH).

Reduction Addition of one or more hydrogen atoms to an organic compound by an enzyme called a REDUCTASE.

Reverse transcriptase An enzyme that makes a DNA copy of an RNA molecule, typically as part of the life cycle of a retrovirus such as HIV.

Ribosome The site of protein synthesis, on which the MESSENGER RNA (mRNA) is TRANSLATED, consisting of structural RNA molecules (the ribosomal RNA) and many kinds of proteins.

RNA polymerase The enzyme that synthesises an RNA copy of a gene, the MESSENGER RNA (mRNA), in the process of TRANSCRIPTION.

Sigma factor The subunit of a bacterial RNA POLYMERASE that gives it specificity for the PROMOTERS of particular genes.

Thymine (T) One of the four BASES making up DNA; it pairs with adenine (A) in the double helix of DNA.

Transcription The process by which RNA POLYMERASE makes MESSENGER RNA (mRNA) from a DNA template.

Transcriptome The complete set of MESSENGER RNA (mRNA) molecules in an organism or cell.

Transduction The process in which a BACTERIOPHAGE picks up bacterial genes from its host and transfers them to a new host, where they are incorporated into its GENOME by CROSSING-OVER.

Transfer RNA (tRNA) A member of the set of RNA molecules needed for TRANSLATION of the MESSENGER RNA (mRNA) on the RIBOSOME by aligning an AMINO ACID with the appropriate CODON on the mRNA.

Transformation The process in which DNA molecules liberated by one bacterium are taken up by another member of the species and incorporated into its GENOME by CROSSING-OVER.

Translation The process by which the MESSENGER RNA (mRNA) directs synthesis of a specific protein on the RIBOSOME.

Transposon A mobile genetic element that can move from one place in a CHROMOSOME, PLASMID, or BACTERIOPHAGE to another; when it inserts into a gene, it usually inactivates the gene's function.

Uracil (U) One of the four BASES making up RNA, substituting for thymine in DNA.

Vector In genetic engineering, a PLASMID or virus into which DNA is inserted *in vitro* and which is used to introduce the DNA into a new host.

Index

Abbott Laboratories, 33, 35, 37–38, 179–180
Actinomyces, 10–12, 17–18, 22, 98
actinomycetes (*Actinomycetales*)
   as antibiotic producers, 21–22, 28–41
   classification/relationships, 16–18, 52, 59, 61–64
   discovery, 10–16
   'rare', 37
actinomycin, 22, 29, 31, 32, 40
Actinoplanes, 38, 41
actinorhodin, 58
   genetics of biosynthesis, 78, 156–160
   in hybrid antibiotic production, 166–167, 178
   in polyketide biosynthesis, 174–176
Actinovate, 129
adriamycin (doxorubicin), 29, 31, 40
A-factor, 160–164
AIDS. *See* HIV
Alikhanian, Sos, 97–100
Alliance for the Prudent Use of Antibiotics (APUA), 48
Amycolatopsis, 41, 219
Amyes, Sebastian, 48

animal testing, of antibiotics, 34, 43
annotation, of gene sequences, 117–119, 194
anthrax, 15, 50, 128, 143
antibiotics. *See also* individual compounds
   and 29, 40–41
   in animal feed, 47–48
   definition, 21
   discovery, 19–23, 29, 35–39
   genetic determination, 154–164
   hybrid, 166–169
   misuse/resistance, 46–49
   naming, 43
   yield increase, 39, 42–43
anticancer agent, 30–31, 40–41, 44, 169, 176, 227
ants, leaf-cutting, 128–129
archaea, 64–65
armadillo, nine-banded, 217–218, 221
Aspergillus nidulans, 51, 68–70
Augmentin, 47
auxotrophic mutant, 57, 59, 87
avermectin, 29, 31, 35, 40, 154, 180 (*see also* Doramectin)
avoparcin, 29, 40, 48

Bacillus subtilis, 136, 142–143, 153
Baltz, Richard (Dick), 88
BCG vaccine, 220–222
Beadle, George, 51, 54, 57, 82
Beijerinck, Martinus, 10, 11
Bentley, Stephen, 118, 121
benzoisochromanequinones (BIQs), 166–167
Beppu, Teruhiko, 160, 162
Bergen, leprosy hospitals, 13–14
Bibb, Mervyn, 85, 94, 96,142, 174, 189
Biotica, 184, 188
BLAST search, 117, 194
bld genes/mutants, of *S. coelicolor*, 140–144, 154, 160, 196
Bloom, Barry, 212–214
Botany School, Cambridge, 53, 55, 60, 75, 83
Buttner, Mark, 125, 138, 152, 154

calcium-dependent antibiotic (CDA), of *S. coelicolor*, 158, 199
Calmette, Albert, 220
candicidin/*Candida albicans*, 29, 30, 40, 43, 180
Cane, David, 176–177, 183
Catcheside, David, 54–56
cell wall, bacterial, 61–62
  as drug target, 29, 40–41, 152
  growth, 130
  in protoplast formation, 86
cellulose, breakdown, 19, 47, 120, 149
cephalosporin, 29, 40, 50
Chain, Ernst, 19, 47, 67, 70
Challis, Greg, 150, 186, 202
chaplins, 125–127, 138
Chater, Keith, 100–101, 105, 128, 140, 142, 144, 154, 225
Chen, Carton, 108–109, 136
chitin, breakdown, 37, 120, 149
chloramphenicol, 29, 33, 40, 43
cholesterol, 31, 39, 227
clinical trials, of antibiotics, 44
code, genetic, 112–113, 117
coelibactin/coelichelin, 151
Cohen, Stanley (Stan), 81, 90–91, 96, 133, 199
Cohn, Ferdinand, 14–16, 18
Cold Spring Harbor Laboratory, 57, 68, 187, 203

collembola (springtails), in spore dispersal, 128
combinatorial biosynthesis, 175–186
combinatorial chemistry, 50, 165
cosmid clones, 109–111, 116, 119, 208–209
crossing-over, 66
  in bacterial genetics, 71–74, 79, 94, 136
  in gene knock-outs, 208–209
  in parasexual cycle, 69
  in protoplast fusion, 88
cyanobacteria (blue-green algae), 65, 128, 193

Davies, Julian, 46, 188
desferrioxamine, 150–151
dihydrogranaticin/dihydrogranatirhodin, 166–167, 169, 174–176, 178
diphtheria toxin, 223
Diversa Corporation, 188, 190
DNA. *See also* PCR, PFGE, transformation, transposon
  microarrays. *See* transcriptome
  recombinant, 81, 90–94
  repetitive, 119–120
  replication, 30, 31, 40–41, 131–133
  sequencing, 114–116
  structure, 92, 112
DNA polymerase, 30, 132–133
  in DNA sequencing, 114–115
  in PCR, 197–198
Doramectin, 189–190
Dubos, René, 19–20, 26, 217

earthworms, in spore dispersal, 128
Enterococcus, infection by, 49, 189
erythromycin
  action/resistance, 34, 41, 46
  discovery, 29, 33
  genetics of biosynthesis/engineering, 179–183, 185–186
Escherichia coli
  antibiotic resistance, 48
  chromosome, 73, 77–78, 105, 131
  fatty acid synthase, 172, 175
  in functional genomics of *Streptomyces*, 207–208
  gene regulation, 139, 159
  genome sequence, 121, 146, 148, 152–153, 193

mating, 57, 73–75, 101, 134–135
  pathogenic strains, 222
  transformation/gene cloning, 91, 93
  transduction, 72
eukaryotes, 52, 61
  genomes, 119–120, 133
  phylogeny, 64–65

F (fertility) factor, 72–74, 94, 101, 134–136
fatty acid synthase (FAS), 169, 171–172, 175, 223
Feldman, William, 23
fermenter, 44–45, 176
fish farms, antibiotic use in, 48
FK506 (tacrolimus)/FK520. *See* immunosuppression
Fleming, Alexander, 3, 19–20, 33, 86
Florey, Howard, 19, 32
Floss, Heinz, 166–168, 176
Food and Drug Administration (FDA), 23, 45, 189
Frost, Lewis, 54, 56, 58

gamma-butyrolactone. *See* A-factor
gas vacuoles, 128, 144
gentamicin, 29, 38, 40
geosmin, 128, 187
GIM (Genetics of Industrial Microorganisms) conferences, 76, 99, 101
Glauert, Audrey and Richard, 59–61
Global Alliance for TB Drug Development, 224
glycogen, in tissue-specific gene expression, 144
gram staining, 21
gramicidin, 20, 29, 40
Guérin, Camille, 220
guinea pig, as host for TB, 15, 23, 24
Hansen, Armauer, 13–14, 16
Harz, Carl, 12–13, 18
Hayes, William (Bill), 72, 83
Henrici, Arthur, 17–18
Hesketh, Andrew, 202
Himshaw, Corwin, 23
HIV/AIDS, 31, 199, 219, 227
homoserine lactone, as bacterial hormone, 162
Hong, Hee-Jeon, 152

Horinouchi, Sueharu, 160, 162
horizontal gene transfer, 121–122, 158, 203
  (*see also* pathogenicity island)
Hutchinson, Richard (Dick), 174–175, 183
hydrophobin, 138

immunosuppression, 31, 40–41, 169, 184–185, 227
insertion sequence (IS), 73, 74, 136
intron, 119–120
iron uptake. *See* siderophore
isoniazid, 219, 224
Istituto Superiore di Sanità, Rome, 67

Jackson, Marianna, 35
Jacob, François, 73, 77, 139, 159
Jacobs, William (Bill), 214, 223, 224
Jakimowicz, Dagmara, 134
John Innes Centre/Institute, 5, 82–84, 174, 183, 214

kanamycin, 29, 34, 40
Kao, Camilla, 183, 206
kasugamycin, 29, 34, 35, 40
Katz, Leonard, 179, 181, 183
Khokhlov, Alexander, 160
Khosla, Chaitan, 176–178, 183
Kieser, Helen, 83, 86–88, 104–105, 107–111, 166–167
Kieser, Tobias, 94–96, 107, 111, 204, 206, 212
Kinashi, Haruyasu, 85, 109, 158
Kitasato Institute, 34–36, 122, 166
Koch, Robert, 14–16, 222
Kornblatt, Mendel, 9, 10
Kosan Biosciences, 183–185, 188, 190

lambda phage, 72, 100, 101, 109
lantibiotic, 138
Lawlor, Elizabeth, 140–141
lazar hospitals, 215, 216
Leadlay, Peter, 177, 179, 183, 184
Lechevalier, Hubert, 30, 34
Lederberg, Joshua, 57, 58, 71–73, 98
Lehmann, Jorgen, 23
leprosy. *See also Mycobacterium leprae*
  discovery of cause, 13–14
  disease process and treatment, 215, 217, 219
  protection, by BCG vaccine, 220

leupeptin, 138
Levy, Stuart, 48–49
light organs, of fish and molluscs (squid), 161–162
Lipman, Jacob, 9–11
Lomovskaya, Natalia (Natasha), 97–101, 105
Losick, Richard (Rich), 136, 140, 142
Luedemann, George, 37–38
lumpy jaw, in cattle, 11–12, 18

malaria, 50, 67, 211
Malpartida, Francisco (Paco), 156, 158, 160, 166, 174–175
Maxygen, in gene shuffling, 90, 189
medermycin/medermhodin, 166–169
membrane, structure, 146–147
mendelian genetics, in USSR, 97–98
Merck, 22–23, 25, 31, 33, 36
messenger RNA. *See* mRNA
methane, in rumen, 47
Microbial Genetics Bulletin, 57
Micromonospora, 18, 37–38, 40
monensin, 29, 41, 47
Monod, Jacques, 139, 159
mRNA (messenger RNA), 29–30 (See also operon, transcription)
  in S_1 nuclease mapping, 194–195
  in transcriptomics, 196–199
MRSA, 49, 203
mutagen(esis), 42, 90, 189–190
mycelium, 16, 61
  aerial, 123–125, 137–144, 154
  vegetative/substrate, 123–125, 129–131, 134, 136, 138, 144, 154
Mycobacterium bovis, 220, 221
Mycobacterium leprae. *See also* leprosy
  discovery, 13–14
  genome sequence, 221, 223
  growth, 217
  oxidative killing, 196
Mycobacterium tuberculosis. *See also* tuberculosis
  discovery, 15–16
  drug resistance, 27, 218–219, 225
  genome sequence, 121–122, 146, 148, 152, 221–224
  growth, 217
  oxidative killing, 196
Mycostop, 129

Nath, Indira, 212
nematode, 31, 154
Neurospora crassa, 51, 53, 57, 82
Nocardia, 18, 219
novobiocin, 29, 33, 41
nystatin, 29, 30, 41

Okazaki fragments, in DNA replication, 132–133
Ōmura, Satoshi, 34, 36, 39, 166–168
operator/operon, 139, 159
oriT, in plasmid mobilization, 74, 204, 207–209
oxidative stress, 153–154, 194, 196

para-amino salicylic acid (PAS), 23
parasexual cycle, 69–70, 75, 86
partitioning, of chromosomes, 133–134
Pasteur Institutes, 73, 101, 139, 220, 221
pathogenicity island, 222–223
PCR (polymerase chain reaction), 197–198
  in gene disruption, 205–209
  in S_1 nuclease mapping, 194
  in TB diagnosis, 223
penicillin
  action/resistance, 29, 41, 47, 49, 53
  chemical synthesis, 33
  discovery, 3, 19–21, 29
  production, 28, 33–34, 42–44, 67, 68, 98
  semisynthetic derivatives, 32, 50
Penicillium, 40–41, 68–70
PFGE (pulsed field gel electrophoresis), 103–108
phage, 62–63, 71–72, 96–98 (See also lambda, phiC31, transduction)
  in fermentation tanks, 44
  for *Mycobacterium*, 214
pharmaceutical companies, 33–34, 47, 76, 88, 90, 156, 189, 227 (See also Abbott, Merck, Schering.)
pharmacokinetics, of antibiotics, 32, 34, 39
phiC31 phage, 97–101
phosphate, as nutrient, 39
phosphate group
  in antibiotic resistance, 46
  in gene regulation, 152–154, 163, 202
pocks, caused by *Streptomyces* plasmids, 94–95, 136

polyethylene glycol (PEG), 86–88, 91, 94, 214
polyketides, 169–183 (*See also* combinatorial biosynthesis)
polymerase chain reaction. *See* PCR
potassium transporter, 148
Pontecorvo, Guido (Ponte), 5 (*See also* parasexual cycle)
   early life 67–70
   in Glasgow, 75–78
   and John Innes Institute, 83, 86–88
potato scab, 223
pristinamycin, 29, 41, 189–190
programmed cell death, 138
prokaryotes, definition, 52, 61
promoter, 139–140, 159 (*See also* Sigma-R)
   and A-factor cascade, 163–164
   of chitinase gene, 149
   for strong expression, 189, 191
protein transport, across membranes, 148
proteome, 199–203
pSAM2 plasmid, 96, 189
Pseudomonas, as pathogens, 49, 162
pulsed field gel electrophoresis. *See* PFGE

quinolones/ciprofloxacin, 50, 225
quorum-sensing, 162

red antibiotic, of *S. coelicolor*, 123, 158–160
Redenbach, Matthias, 109–110
replica plating, 58
restriction enzyme, 93
   in genetic engineering, 90, 93
   in PFGE, 104–108
reverse transcription, 31, 199
rhizosphere, 129
ribosome, 29, 30, 64, 118
   as drug target, 34, 40–41, 46, 185, 219
ribosome binding site, 117
rifamycin/rifampicin, 29, 41, 219
river blindness, 31
RNA polymerase, 30, 139, 219 (*See also* sigma factor)
Rockefeller Institute/University, 19–20, 57, 71, 148
Roper, Alan, 69–70, 76
rumen, 47
Rutgers College/University, 9–11, 18–19, 23, 26

S_1 nuclease mapping, 194–196
Saccharopolyspora erythraea, 40, 180, 183, 186
Salmonella, 21, 52, 71–72
Salvarsan, 11, 21, 165
Sanger Centre/Institute, 110–112, 114, 116, 119, 221, 224
Sanger, Fred, 113–115
Schatz, Albert, 23, 26
Schering Corporation. 37–38
Schrempf, Hildgund, 85, 148–151
SCP1 plasmid, 84–85, 87, 119, 136, 157
SCP2 plasmid, 85, 87, 94, 96, 119
secondary metabolite, 187–188, 202, 227
Sermonti, Giuseppe, 64, 66–67, 70, 75–77
Sherman, David, 173–176
siderophore, 150–151, 187
sigma factor, 139–140, 151–153
   ECF family, 153, 207, 209
   Sigma-R, 154–155, 194, 196
   in *Streptomyces* and *Bacillus* development, 142–144
SLP1 plasmid, 94, 96
Southern blotting, 107
Spores, of *Streptomyces*
   development, 142–144
   dormancy/dispersal, 124–129
Staphylococcus, 21, 119 (*See also* MRSA)
starch, as microbial food source, 120, 149
statins/lovastatin, 29, 31, 41, 227
Streptomyces aureofaciens, producer of tetracycline antibiotics, 40–41, 43, 98
Streptomyces avermitilis, avermectin producer/genome sequence, 40, 122, 144, 187
Streptomyces coelicolor, 4, 58
   antibiotics. *See* actinorhodin, calcium dependent antibiotic, red antibiotic
   developmental biology, 140–144 (*See also bld* genes, *whi* genes)
   genetic map, 77–78, 103–111
   genome sequence, 116–122, 125, 144, 146,148–152
   host for unnatural natural products, 178, 183, 185
   mating/plasmids, 72–73, 136 (*See also* SCP1, SCP2, SLP1)
   protoplasts, 86–87, 89, 91, 94
Streptomyces diversa, as host for unnatural natural products, 191

Streptomyces griseus, streptomycin producer, 23, 26–27, 41, 46, 159–161, 163
Streptomyces lividans 100 (*See also* potassium transporter, SLP1 plasmid)
 antibiotic sensitive mutant, 225
 linear chromosome, 108
 SLP2 plasmid, 121
Streptomyces rochei, plasmid-determined antibiotic production, 158
Streptomyces virginiae, producer of virginiamycin and gamma-butyrolactones, 41, 161
streptomycin
 action/resistance, 23, 29, 34, 41
 discovery, 23, 26, 29
 effect on TB incidence, 26, 217–219
 genetics of biosynthesis, 159–160, 163–164
streptothricin, 22–23, 29, 41, 43
Streptothrix foesteri, 14–16, 18, 22
Stutzman-Engwall, Kim, 189, 206
sulphonamides, 21, 50, 165
Swann Report, 48
Synercid, 189–190

Takano, Eriko, 141, 164
Tatum, Edward, 51, 54, 57, 72, 82
TDR programme, 211–212
teicoplanin, 29, 38, 41, 227
terminal inverted repeat (TIR), 109, 137, 204
tetracenomycin, 174–175, 178–179
tetracycline antibiotics
 action/resistance, 34, 40–41, 46, 147
 in animal feed, 47
 discovery, 29, 33–34
Thompson, Charles, 96, 202, 225
tissue-specific gene expression, 144, 199
transcription, as drug target, 29–31, 40
transcriptome, 196–199
transduction, 52, 71–72
transformation, 51, 71–72
 of *Escherichia coli*, 91, 208
 of *Streptomyces*, 94
translation, as drug target, 29–30
transporter, membrane, 146–150, 191
transposon, 122, 203–207
tRNA (transfer RNA), *See bldA*
trypsin, 138, 200–201

tuberculosis (TB). *See also* BCG vaccine, *Mycobacterium tuberculosis*
 diagnosis, 223
 discovery of cause, 15
 disease process/latency, 215–218
 incidence, 26, 215, 217
 treatment, 23–26, 27, 218–219, 224–225, 227
two-component regulator, 152–153
tylosin, 29, 41, 90
tyrocidine/tyrothricin, 20

Umezawa, Hamao, 34, 35
undecylprodigiosin. *See* red antibiotic
University of East Anglia (UEA), 5, 82–83
University of Glasgow, 5, 68, 70, 75–77
urinary tract infection, 21, 222

vancomycin 29
 action, 32, 41
 in microscopic staining, 130
 resistance, 48–49, 151–152, 189
Vibrio, 21, 161–162
virginamycin, 29, 41, 161
Vivian, Alan, 83, 84

Waksman, Selman, 20, 25, 55
 classification of actinomycetes, 17–18, 37, 63
 discovery of antibiotics, 19–23, 28
 early life, 8–11, 16, 98
 Nobel Prize controversy, 23, 26
Waksman Institute, 23, 25, 64
Wellcome Trust, 110, 114, 119
Westpheling, Janet (Jan), 85, 140, 149, 181
whi genes/mutants, of *S. coelicolor*, 140, 142–144, 225
Whitehouse, Harold, 53–55, 64
Winogradsky, Sergei, 10, 11, 19
Woese, Carl, 63–64
Woodruff, Boyd, 22, 36
World Health Organization (WHO), 48, 211, 215
Woyceisjes, Amerigo, 37–38
Wrocław (Breslau), 14–15, 131, 134

X-rays, as mutagens, 42, 68, 187

Zakrzewska-Czerwinska, Jolanta, 131, 134